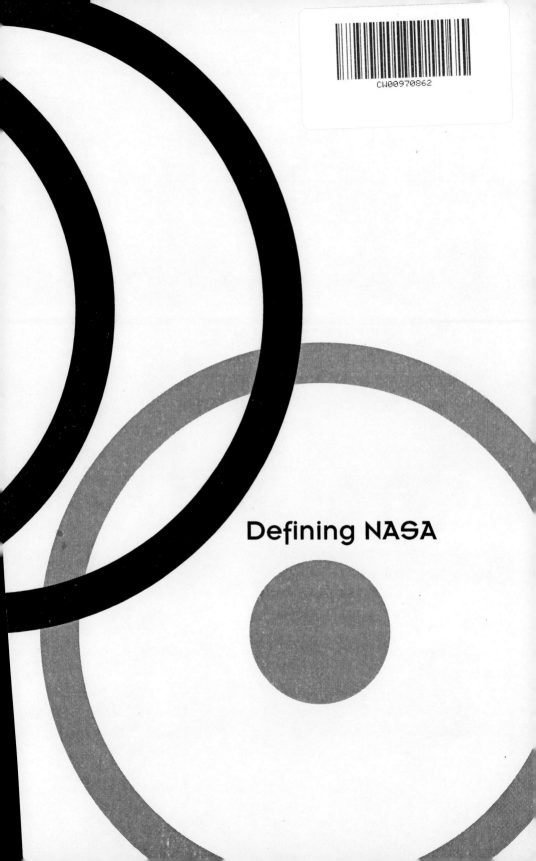
Defining NASA

Defining NASA

The Historical Debate over the Agency's Mission

W. D. Kay

State University of New York Press

Published by
State University of New York Press, Albany

© 2005 State University of New York

All rights reserved

Printed in the United States of America

No part of this book may be used or reproduced in any manner whatsoever without written permission. No part of this book may be stored in a retrieval system or transmitted in any form or by any means including electronic, electrostatic, magnetic tape, mechanical, photocopying, recording, or otherwise without the prior permission in writing of the publisher.

For information, address State University of New York Press,
194 Washington Avenue, Suite 305, Albany, NY 12210-2365

Production by Kelli Williams
Marketing by Michael Campochiaro

Library of Congress Cataloging-in-Publication Data

Kay, W. D.
 Defining NASA : the historical debate over the agency's mission / W. D. Kay.
   p. cm
 Includes bibliographical references and index.
   ISBN 0-7914-6381-8 (hardcover: alk. paper) — 0-7914-6382-6 (pbk.: alk. paper)
 1. United States. National Aeronautics and Space Administration. 2. Space sciences—United States—History. 3. Astronautics—United States—History. I. Title.
TL521.312.K38 2005
354.79'0973—dc22

2004009798

10 9 8 7 6 5 4 3 2 1

To Melanie and Will, who were, in so many ways, an integral part of the process

Contents

Acknowledgments ix
List of Abbreviations xi

Part One: Introduction 1
Chapter 1 What Is NASA's Purpose? 3
Chapter 2 Analytical Framework 11

Part Two: First Mission 25
Chapter 3 Prehistory: Space Policy Before *Sputnik* 27
Chapter 4 NASA: Born Out of Fright (1957–1961) 41
Chapter 5 Mission Advanced 67

Part Three: Second Mission 89
Chapter 6 Mission Accomplished . . . Now What? 91
Chapter 7 Space Policy Redefined (Again) 115
Chapter 8 Dollars, Not Dreams; Business, Not Government 143
Chapter 9 Concluding Thoughts 165
Notes 179
Index 243

Acknowledgments

Like some other NASA programs one could name, this project took just a tad longer than planned, and looks rather different from what its designers envisioned. Way back in 1996, the agency's History Office commissioned a manuscript, tentatively titled *Contested Ground: The Historical Debate Over NASA's Mission*, that would provide an account of the different goals and objectives that had been set for the agency—as well as those that it had sought on its own—from 1958 to the present. I, however, being a political scientist and all, proceeded to alter the prescribed design to include a bit of policy analysis here and there. His has resulted in a narrative that a mainstream historian is apt to regard as a tad "wonky," and a true policy wonk will see as dated (the public policy analyst's word for historical).

Thus, first and foremost, I need to thank the NASA History Office in Washington, DC, and particularly the former chief historian, Roger Launius—now at the Smithsonian—for his support and his patience over a few more years than either of us expected. The staff (some of whom have moved on)—Lee D. Saegesser, J. D. Hunley, Steve Garber, Nadine Andreassen—were, as always, a joy to work with. It was also through NASA that I encountered Andrew Butrica, a great scholar who was always willing to provide aid, advice, and encouragement.

I acknowledged NASA first, because, well, they supported the research. It is my family, however, who deserves, my deepest thanks. Given all she has had to put up with, that really ought to be a picture of my wife, Jen, on the cover instead of Ed White (that would probably have helped with sales, too). Absent that, there are no words for how much I owe her, except: she is absolutely the love of my life. And speaking of life, there are two individuals who did not even exist when this project began. I vividly recall the tiny baby sleeping on my desk when I first began working on the text. Melanie is now all of seven years old, and has literally grown up alongside this book. Her younger brother, Will, has as well, although at four years of age, he is largely oblivious. As I said in the dedication, they both were every bit a part of this.

The historical data comes from all across the country, and there are literally dozens of people who were instrumental in helping me collect it. The staff members of the Kennedy, Johnson, Carter, and Reagan Presidential Libraries, along with their colleagues at the National Archives in Maryland, could not have been more helpful. Given the later chapters' discussion of space commerce, the book would have been even more difficult to write had it not been for the wonderful folks at the Space Business Archives, who let me rustle about in their collection while it was still being pulled together. It was also Melanie's first long car trip.

Happily, I was once again able to draw upon the insight and wisdom of my colleagues in the Organized Section on Science, Technology, and Environmental Politics of the American Political Science Association, especially Pat Hamlett, Ned Woodhouse, Gary Bryner, Frank Laird, Dave Guston, Harry Lambright, Vicki Golich, Steve Collins, and, most particularly, the section chair, J. P. Singh. It was J. P. who introduced me to Michael Rinella of State University of New York Press, to whom, along with Kelli Williams, I also owe a huge debt.

Finally, I have been very fortunate to have spent the last decade and a half among a group of academics in which the level of collegiality borders on the surreal. The Department of Political Science at Northeastern University is stocked with some of the finest minds I've ever known: Bob Gilbert, John Portz, Denis Sullivan, David Rochefort, and the masterful Chris Bosso all made important contributions of one kind of another. Nor can I forget the tireless efforts of Shannon MacMahon, the greatest graduate assistant in the world.

For those countless others—congressional staffers, the folks at the National Science Foundation, the National Institutes of Health, the Department of Energy, and any number of foundations, academic departments, and university libraries—who I have neglected to mention by name: Please do not think me ungrateful—as Jen so often tells me, my head is built like a sieve. That fact, incidentally, also accounts, for whatever errors remain in the text. I can at least take some comfort from the fact that this time I managed to avoid making any *Star Trek* references—those are nothing but trouble.

Abbreviations

AEC	Atomic Energy Commission
AIDS	Acquired Immunodeficiency Syndrome
ARPA	Advanced Research Projects Agency
ASAT	Anti-Satellite Weapon
ATF	Bureau of Alcohol, Tobacco, and Firearms
BMD	Ballistic Missile Defense
Caltech	California Institute of Technology
CCCT	Cabinet Council on Commerce and Trade
CCDS	Centers for the Commercial Development of Space
CDSF	Commercially Developed Space Facility
CIA	Central Intelligence Agency
CRADA	Cooperative Research and Development Agreement
DARPA	Defense Advanced Research Projects Agency
DOC	Department of Commerce
DOD	Department of Defense
DOE	Department of Energy
DOT	Department of Transportation
EOR	Earth Orbit Rendezvous
EOSAT	Earth Observation Satellites Corporation
ESA	European Space Agency
FAA	Federal Aviation Administration
FCC	Federal Communications Commission
FY	Fiscal Year
GeV	Billion Electron Volt
GOP	Republican Party
HHS	Department of Health and Human Services
ICBM	Intercontinental Ballistic Missile
IEEE	Institute of Electrical and Electronics Engineers
IGY	International Geophysical Year
ISF	Industrial Space Facility
LOR	lunar orbit rendezvous
NASA	National Aeronautics and Space Administration

NRL	Naval Research Laboratory
NSC	National Security Council
NSD	National Security Directive
NSDD	National Security Decision Directive
NSF	National Science Foundation
OMB	Office of Management and Budget
OSTP	Office of Science and Technology Policy
PSAC	President's Science Advisory Council
R&D	Research and Development
RCA	Radio Corporation of America
SDI	Strategic Defense Initiative
SIG (Space)	Senior Interagency Group for Space
SII	Space Industries Incorporated
SSC	Superconducting Super Collider
SSI	Space Services Inc.
STG	Space Task Group
STS	Space Transportation System
USSR	Union of Soviet Socialist Republics
VfR	German Society for Spaceship Travel

PART I

INTRODUCTION

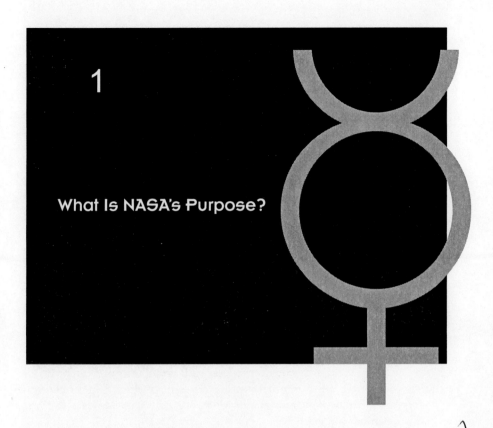

1
What Is NASA's Purpose?

Since the opening of the Space Age nearly 50 years ago, the U.S. government has spent close to $1 *trillion* on space-related activities.[1] Although by no means the largest public expenditure over that period—it is easily dwarfed by national defense, social security, and countless other federal programs—it is still, by any measure, an enormous amount of money. It therefore seems perfectly reasonable to ask: what has the American taxpayer gotten in return for this massive investment?

Most space advocates, of course, will immediately point to the myriad benefits that flow from space-based research and development. They would note the advances and improvements in weather forecasting, communications, and navigation, and argue that the economic impact of these systems alone—which have accounted for the creation of not only thousands of jobs, but of dozens of whole new industries as well—probably exceeds $1 trillion (it is certainly in the hundreds of billions). They might also claim that satellite and rocket technologies have made a contribution to American national security that is beyond economic calculation.

And these are just the direct applications. Although critics often scoff at the supposedly inflated claims made for the "spin-offs" of the space program,[2] there is no denying that a large number of commercial products, services, and technologies in common use today were originally developed for space missions. It is, of course, impossible to determine with any degree of precision the total dollar value of these indirect benefits, but they are certainly not insignificant.[3]

None of these assertions, however, completely answers the question. After all, very few people seriously argue that the United States should have no space program whatsoever. Actually, most—if not all—of the controversy regarding U.S. space policy over the past half century has revolved around the question of how to proceed: what should the U.S. be doing in space, and when?[4] Indeed, the fact that space represents such an enormously valuable resource only underscores the importance of that $1 trillion investment. Given what is at stake, can the taxpayers really be assured that these very large sums they have entrusted to their government have always been spent in the most productive fashion?

Inevitably, it is the operation and programs of the National Aeronautics and Space Administration (NASA) that becomes the focus of such questions. Although far from the only (or even, at times, the largest) federal agency engaged in space-related activities,[5] for more than 40 years, NASA has stood as the country's most visible, and most important, public space organization. Thus, it is NASA, more than any other agency or department, that has been confronted with the vexing question, "Why?"

During the early 1960s, opponents of Project Mercury questioned the rush to place a man in space. A few years later, skeptics were disputing the value of the Apollo moon program. Shortly after that, NASA faced stiff opposition over the development of the space shuttle, and, still later, had to fend off—for a time, on an annual basis—policymakers determined to cancel the International Space Station. In addition, there are a number of cases, such as the post-Apollo proposals for a mission to Mars, the National Aerospace Plane, and the first Bush administration's Space Exploration Initiative, in which proponents could not answer the "why" questions convincingly.[6]

For many critics, however, the problem lies not so much in the value of any specific project as in the agency's overall strategy for exploring and using space. By far the largest of these debates involves NASA's commitment to sending people into space. Opponents of the agency's human spaceflight program have long argued that anything that humans can do in space can almost always be done more easily, at far lower cost, and with much less personal risk by unmanned rockets and automated probes. The fact that this argument has been raging unabated for nearly 50 years, dogging every major project from Mercury to the International Space Station, suggests that it will not be resolved anytime soon.

Another controversy that emerged during the 1980s concerns the agency's role in space "operations." A number of observers, particularly economic and political conservatives, complain that NASA is too involved with providing "space services," a task that is not only inconsistent with its "true" mission as a research and development (R&D) organization, but is also impeding the development of private sector space activities. This particular debate has evolved over the past two decades into a more general discussion about the role of government in space exploration overall.

Of course, there can be no doubt that some of the criticism leveled at NASA over the past 45 years has been fully warranted. Like any human enterprise, the agency has clearly made mistakes, some of which were quite serious (and even lethal).

Moreover, even if it had been operating perfectly (whatever that might mean), any organization that spends billions of the public's dollars every year has to expect to displease someone. Finally, because it combines the elements of very large costs, high visibility, and extreme risk—few public programs face the possibility of failure on the scale of the Mars Polar Lander or the *Challenger* and the *Columbia* shuttles—spaceflight is very likely an area of public policy that will always invite controversy.[7]

One of the major themes of this book is that almost all of the controversies surrounding NASA—its priorities, its management, even its occasional (or, as some of its most vocal critics would have it, its frequent) errors—have a common root. There is one fundamental question about the agency that has seldom even been asked directly, let alone examined in any detail. Stated simply: at no point in NASA's history has there ever been a clear, specific statement of its actual purpose. What, in a word, is the agency is *for*?

Of course, asking this question immediately gives rise to another: what is it for, according to whom? NASA officials, obviously, can provide a highly detailed account of what they do and, from their perspective, its value to society. Indeed, they do so every year during budget hearings. Moreover, current management practice—not to mention federal law—requires that the agency publish a "mission statement" that, among other things, lays out its overall purpose (more about this below).

Unfortunately, declarations of this sort cannot (at least not by themselves) provide a completely satisfactory answer to the question that opened this chapter. Every year, for nearly half a century, presidents and their appointees, working in conjunction with the elected members of Congress, have chosen to spend billions of dollars on space instead of defense, education, health care, welfare, or any number of other pressing public needs. Presumably, they have done this not for NASA's comfort and convenience, nor simply to provide work for its employees (although critics charge that it seems that way at times), but because they have believed—we will assume sincerely—that it would in some way promote the public interest. Thus, the issue at hand should more properly be phrased: what does the government—the public's representatives and the guardians of their interests—think that the agency is for?

As might be expected, there cannot be a single answer to this question. During the latter half of the 20th century, conditions in the country and in the world changed considerably, and government priorities, as a matter of course, changed along with them. Since NASA was founded, there have been 22 years of Republican presidents, 16 of Democratic. Thousands of members of Congress have come and gone. There have been recessions, periods of high inflation, and some years of great prosperity. Last, but far from least, the Cold War raged, subsided, and ended altogether. It is therefore only reasonable that succeeding governments spread across nearly 50 years would hold differing views about the U.S. space program in general and NASA's role in particular.

The purpose of this book is to examine those changing views. Drawing on relevant concepts in political science and public administration, it attempts to account for the shifts in U.S. space policy, and to describe the political, economic, and technical factors that helped bring them about. Among the questions to be addressed are: what did policymakers envision when they created the agency in the first place? Why was there such strong (albeit not entirely unanimous) support for Project Apollo in the early 1960s, and why did it fade away in such a relatively short time? What caused NASA and the program to languish throughout most of the 1970s, and why did it reemerge (after a fashion) in the 1980s? Finally, what role is the agency expected to play today?

Original Intent versus "What's in It for Me?"

The most obvious place to begin looking for NASA's purpose would seem to be the legislation that created it, the National Aeronautics and Space Act of 1958. Unfortunately, that document provides very little in the way of a practical guide to the agency's activities. To begin with, the description of its responsibilities is, to put it mildly, somewhat broad:

- The expansion of human knowledge of the phenomena in the atmosphere and space.

- The improvement of the usefulness, performance, speed, safety, and efficiency of aeronautical and space vehicles.

- The development and operation of vehicles capable of carrying instruments, equipment, supplies, and living organisms through space.

- Conducting "long-range studies" of the potential benefits, opportunities, and problems of aeronautical and space activities.

- The preservation of the United States as a leader in aeronautical and space science and technology.

- Providing relevant information to DOD and related agencies.

- Cooperation with other nations in the peaceful pursuit of aeronautical and space activities.

- Cooperation and coordination with other public agencies to avoid "unnecessary duplication of effort, facilities, and equipment."[8]

Second, the act makes no effort to assign any sort of priorities to these tasks. Read literally, it appears to give equal weight to the "carrying" of instruments, equipment, supplies, and living organisms (which presumably includes humans) into space. Finally, apart from allowing the agency to enter into contracts with industry and educational institutions[9] and calling for the "widest possible practicable and appropriate dissemination of information"[10] (which appears to refer to commercial applications as well as scientific research findings), the legislation does not seem to envision any significant role for the private sector.[11] In short, the Space Act, for all of its historical importance, really does not help answer any of the current controversies concerning NASA and its operations. Like most other pieces of authorizing legislation, it simply provides the broad contours in which the agency must function.

There are similar problems in NASA's own "vision" and "mission" statements, which, as noted earlier, are now included in all of its public documents. The agency's 2000 Strategic Plan, for example, describes as its vision:

> NASA is an investment in America's future. As explorers, pioneers, and innovators, we boldly expand the frontiers in air and space to inspire and serve America and to benefit the quality of life on Earth.

And its mission as:

- To advance and communicate scientific knowledge and understanding of the Earth, the solar system, and the universe.

- To advance human exploration, use, and development of space.

- To research, develop, verify, and transfer advance aeronautics and space technologies.[12]

Although they do make a somewhat more direct acknowledgment of the private sector (such as the references to "innovators" and technology transfer), these statements are just as broadly cast as those of the Space Act. Once again, this is not terribly surprising, inasmuch as that is not really the purpose of a mission statement. It is simply supposed to describe what might be called an organization's "operating philosophy." Despite the name, it cannot—nor is it intended to—provide a precise account of NASA's mission, as that term is being used here.

In actual practice, the agency's long-range goals, priorities, timetables, and even to some degree its method of operation (as with any major public organization) are shaped by a number of complex factors, including the bureaucratic, budgetary, legislative, electoral, and other political processes of the U.S. federal government, as well as the demands of a wide variety of outside interest groups.[13] In addition, NASA's internal procedures and relationships, its core organizational values, and its view of external events (which includes its understanding of its political, economic, and social environment) are all the product of a rich and diverse organizational culture that has been evolving for nearly a century.[14] The agency is also at the mercy of a whole range of impersonal forces, such as the state of the economy, the pace of scientific discovery and technological development, unplanned events (e.g., the Apollo fire or the *Challenger* and *Columbia* accidents), and—increasingly—international affairs. Making matters even more complicated is the fact that these are all subject to sudden and, for all practical purposes, unpredictable shifts.

Finally, there are the vagaries of Washington politics. NASA does, after all, exist in a highly charged political environment. This can be most clearly seen in the annual appropriations battles, where, under the peculiar structure that makes up the congressional budget-writing process, the agency is required to compete head to head for its funding with completely unrelated organizations, such as the Department of Housing and Urban Development and the Department of Veterans' Affairs. Needless to say, this odd arrangement has led to some unusual political trade-offs over the years.¹⁵ Finally, NASA has, on occasion, found itself at the center of a political party's—or even an individual policymaker's—ambitions. The impact of this vast array of political, economic, and social forces cannot possibly be captured in formal statements.

Perhaps, then, a better way to approach the question of NASA's mission is simply to ask: what could the agency be for? What are the public benefits that supposedly flow from space-related activities? A list of such benefits would include (in no particular order):

- *Scientific research.* It is difficult to think of an area of basic science that has not been affected by the development of space technologies. Astronomers in particular have benefited from planetary and deep-space probes, as well as automated observatories, like the Hubble telescope, that orbit above the earth's distorting atmosphere. Geologists, geophysicists, and hydrologists make extensive use of remote-sensing satellites. Useful research in biology, chemistry, and physics can be conducted on board orbital facilities.¹⁶

- *Economic and commercial applications; "making life better here on earth."* As suggested earlier, the advances in weather forecasting made possible by meteorological satellites have led to improvements in such areas as disaster preparedness and agricultural planning, just as the growing number of relay satellites has opened up a vast—and still expanding—global network of radio, television, telephone, and Internet communications. Remote-sensing satellites—which can be used to detect oil, gas, and mineral deposits, as well as to monitor pollution, deforestation, and other changes in the earth's environment—have opened up a whole host of new business opportunities, as has a global network of navigation satellites. Last, but not least, many enthusiasts still hold out hope for space-based manufacturing, which could (in theory) lead to the creation of new metals, medicines, and other useful products.¹⁷

- *Intellectual stimulation and discovery; satisfying the "urge to explore.* It is often said that human beings (and particularly Americans) have an innate desire to understand the unknown. According to this view, it is the mark of a "great nation" to expend some of its resources in meeting this challenge. Space advocates also like to point to the "changes in perspective" that can come from "conquering space," such as viewing the "whole earth" (i.e., a world without visible national boundaries) from orbit or the surface of the moon, or of finding life on other planets.¹⁸

- *National defense (narrowly conceived).* Obviously, rocket technology (in the form of guided missiles) has direct military application. In addition, the armed forces make extensive use of navigation, communication, and surveillance satellites. Thus, although international treaties ban weapon systems themselves from space, military planners still regard it as a sort of strategic "high ground."

- *National security (broadly conceived; "national prestige").* Success in space, according to some, is the most direct and dramatic demonstration of a country's talents and abilities (and, by implication, its power). On October 15, 2003, the People's Republic of China became the most recent example of this. Yang LiWei's 14 orbits of the earth has underscored China's status as a rising superpower.[19] Proponents of such a program for Japan have a similar aim.[20]

What is remarkable about this list is how little it has changed over the last 50 years, despite significant developments in both the relevant technologies and world politics. Enthusiasts were extolling the virtues of communication and weather satellites long before such applications were even remotely feasible (in fact, the earliest articles on radio and television relay satellites appeared even before the end of World War II). Similarly, as late as the mid-1980s, long after the United States had "won" the "space race," supporters of the space station were painting a vivid picture of the Soviet space station program in an obvious (but ultimately unsuccessful) attempt to appeal to American national pride.

While the contents of the list may have stayed the same, the relative political saliency of the items—that is, the degree of importance attached to them by policymakers—has changed quite dramatically. Forty years ago, fear that it was losing its "prestige" as a world power was enough to push the United States—literally by itself—into the most ambitious (and most costly) space venture in history. Just a few years later, however, such concerns had largely evaporated. The efforts of the station's proponents notwithstanding, there has not been a single major NASA program initiated since the mid-1960s that can really be said to have been motivated by prestige. Clearly, then, the key to understanding the changing nature of the agency's mission—and the primary approach of the book—will be to determine exactly which benefits of space technology can be said to be "driving" the program at any given point in time.[21]

The Plan of This Book

Along with providing some new insight into the ups and downs of the U.S. space program, it is hoped that this book will provide a fresh view of a few other space-related subjects as well. First, as numerous historians have noted, there is a tendency in much of the current literature to overemphasize the role of the American president in setting the course of the space program.[22] Obviously, some presidents have played key roles in the history of the program, and at times—John F. Kennedy with Project Apollo and Ronald Reagan with the space station program—have been the central, indeed, the pivotal decision maker. The point, rather, is that most of the time that individual is only one actor within a much larger political system. For this reason, it is also important to take into account the actions of other governing institutions—most notably, the U.S. Congress—in setting space policy. Thus, while the analysis by no means ignores the actions of presidents (in fact, the research was conducted at four presidential libraries), every effort will be made to view space as a government—not just a presidential—program.

Second, the concepts that will be used here suggest a different interpretation of some of the major episodes in space history—including the U.S. response to *Sputnik*, Kennedy's lunar landing declaration, the decision to develop the space shuttle, and Reagan's approval of the space station—than that found in some of the existing literature.[23] They also shed light on some issues—Kennedy's proposals for a joint U.S.–USSR moon flight, NASA's post-Apollo decline, and the reformulation of space policy in the 1980s—that have not received as much attention.

Finally, this book attempts to place the history of the space program within the larger context of overall postwar U.S. R&D policy. Although many scientists and engineers have traditionally viewed NASA as a competitor of sorts (and a greedy one at that), it can be shown that many of the same political, economic, and social forces that shaped the course of space policy had a nearly identical impact on most other publicly funded science and technology programs. Indeed, as will be discussed in the concluding chapter, the experiences of NASA after the Apollo era actually provided a preview of sorts of many of the problems facing U.S. R&D policy today.

The argument proceeds as follows. Chapter 2 concludes this introductory section by laying out several concepts from the literature in political science, public administration, and public policy studies that will guide the remainder of the analysis. The three chapters comprising part 2 then apply these concepts to the early history of spaceflight in the United States (and, to a lesser extent, elsewhere), starting with the earliest musings about space and the beginning of rocket research (chapter 3) and continuing through the end of the so-called Golden Age of U.S. space policy (chapters 4 and 5). As will be seen, one of the major factors contributing to NASA's success during this period is that it was the first—and so far the only—time that a common agreement existed among the relevant stakeholders as to the nature of NASA's mission.

The four chapters comprising part 3 examine NASA's struggle to find a new purpose following the Apollo moon landings.[24] Chapter 6 describes the "malaise" in the U.S. space program during the 1970s. Chapter 7 looks at the political changes of the early 1980s, which brought a group of policymakers to Washington who were determined to move space policy in a completely new direction. Chapter 8 follows these developments through the end of the 1990s. Finally, chapter 9 assesses the status of NASA, the U.S. space program (or, as will be seen, *programs*), and American R&D policy in general entering the 21st century.

Unfortunately, and at the risk of making the reader feel cheated, one thing this analysis cannot do is answer the question posed at the beginning, that is, whether the taxpayers' $1 trillion investment in space has been wisely spent. This is not a matter that can be settled objectively. What the following chapters hopefully will do, however, is provide a better understanding of the political, economic, and technical factors guiding policymakers as they decided which space investments were worthwhile. Such an understanding, in turn, may help settle a critical, and much-debated issue: where does the space program—and NASA—go from here?

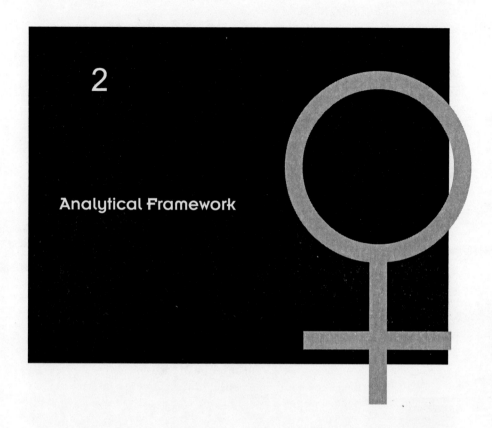

2
Analytical Framework

At first glance, accounting for the origin and evolution of NASA's "mission"—or, indeed, that of any large, complex, big-budget, public organization—would appear to be an exceedingly difficult task. Fortunately, a number of concepts from the fields of political science, public administration, and public policy studies may make such an inquiry, if not easier, at least more systematic. Although only a few have been applied specifically to the space program, these ideas have proven useful in examining other U.S. science and technology programs, as well as other areas of public policy generally. Taken together, they should provide some insight into the origin and development of NASA's mission(s).

Problem Definition

Sociologist David Dery tells a story about a hotel manager who must deal with complaints from guests about long waits for the elevators. An engineering consultant has recommended either putting in additional elevators or installing equipment to make the existing ones move faster. A psychologist, on the other hand, suggests placing mirrors or some sort of interesting or informative items along the walls next to the elevators. What should the manager do?[1] Clearly, making the "correct" choice depends on what kind of difficulty—technical or psychological—the manager believes the hotel is facing. In other words, a crucial first step in addressing the complaints would be to define the problem at hand.

As chapter 1 noted, there is no shortage of objectives for a national space program: economic growth, technological development, defense, national prestige, scientific research, and so forth. Determining which of these have been (or, for that matter, ought to be) the primary focus of NASA's mission, however, invites a prior question rather similar to the hotel manager's: what is/are "the problem(s)" in space that policymakers expect NASA to address?

This issue is by no means unique to space policy (nor, for that matter, to hotel management). All public issues of any degree of complexity cut across a number of established policy fields. Tobacco consumption is simultaneously an agricultural, trade, public health, and education issue,[2] and has recently shown up in discussions of product liability, insurance, and workplace safety as well. A firm's decision to close down a plant or factory can be seen as a labor question, a local economic concern, or an element of a company's competitive strategy.[3] Coal mining is both an environmental and an energy issue,[4] and also raises serious questions regarding land use and occupational safety.

Public officials almost never attempt to address each and every facet of such issues, and certainly not all at the same time. Rather, they tend to focus their attention on one or two particularly salient aspects. The War on Drugs of the 1980s defined drug usage as a crime, as opposed to an illness, and thus emphasized law enforcement over—if not to the exclusion of—clinical treatment.[5] The 1974 decision to authorize construction of the Alaska Pipeline was based on concerns over the supply of energy, rather than on environmental criteria.[6]

Reaching a consensus on problem definition can be a difficult and protracted process for public officials. For example, there is still considerable disagreement among policymakers, advocates, and the public at large as to whether homelessness in the United States ought to be viewed as an economic, housing, mental health, or some other type of problem.[7] Similarly, during the initial stages of the AIDS epidemic in the United States and Europe, some activists felt that the disease had been incorrectly labeled as a purely scientific problem, to be addressed primarily by virologists and other medical researchers, rather than a public health/education issue.[8]

As these examples suggest, there is far more at stake in issue definition than semantics or symbolism. How policymakers ultimately choose to define a public problem determines a great deal about how—or even if—they will seek to address it. Indeed, many of the most critical features of the policymaking process are shaped by the manner in which the problem to be tackled is conceptualized.

Problem Ownership

Acceptance of a definition usually (but not always—see below) settles the question of problem ownership, that is, which agency or organization is to acquire jurisdiction over the issue. Initially, defining AIDS as a virological problem made its assignment to the Centers for Disease Control and the National Institutes of Health—as opposed to an educational agency like the Public Health Service— inevitable. In much the same way, the categorization of illegal drugs as a law enforcement issue automatically made it a responsibility of the relevant agencies within the Department of Justice, rather than those at the Department of Health and Human Services.

There are cases, of course, where the definition-ownership relationship runs in the opposite direction, that is, where an agency and its allies attempt to get a problem defined in a particular way precisely in order to gain (or retain) control over some project or program. When the Carter administration split the U.S. Department of Health, Education, and Welfare into the Department of Education and the Department of Health and Human Services (HHS), a serious conflict arose over which one would inherit the politically popular Head Start program. In essence, the argument boiled down to a difference over whether the program was to be considered a part of educational or of social welfare policy (HHS' view ultimately prevailed).[9]

Similarly, the armed services sought to obtain total control over the U.S. atomic energy program in the years immediately following World War II by claiming that such research was primarily (if not exclusively) for the purpose of developing weaponry. Advocates of a civilian program argued for a broader definition, which would include such applications as power generation and propulsion. This is the perspective that won out in the end, with ownership of U.S. nuclear energy policy passing to the civilian Atomic Energy Commission.[i]

Sifting Information

Defining an issue and assigning ownership play an important role in helping policymakers screen out "irrelevant" information, or, more precisely, in helping them determine what types—and sources—of information are to be considered relevant. This is most readily apparent when it comes to evaluating expertise.[11] In Dery's hotel example, the psychologist's opinions become moot the instant the manager decides that the elevators are a mechanical problem (just as, obviously, the engineer's recommendations would not be applicable to a psychological problem).

To take an actual policy case, consider the current controversy over fetal tissue research. Opponents of the practice, the vast majority of whom see it as a religious or moral question, generally pay little attention to expert claims concerning its medical potential (such as treating Parkinson's disease). For those who regard such research as inherently immoral, any assertions about its possible benefits simply has no meaning.[12] Moreover, a number of critics are convinced that the acceptance of fetal tissue use would increase the number of abortions performed in the United States. Put another way, some opponents define this as a pro-life issue, for which, once again, the presumed therapeutic benefits are not germane.[13]

As this last point—the belief that fetal tissue research encourages abortions—suggests, problem definition is very strongly bound up with how government officials, activists, and even the general public perceive the issue's various cause–effect relationships. Much of the conflict over the definition of homelessness in the United States stems from the fact that there is still no widespread agreement as to why some people wind up living on the streets. Differing views on the causes of the 1992 Los Angeles riots—anger over the Rodney King verdict, persistent problems of poverty and racism, corrosive effects of federal welfare programs, decline of respect for the law, and so forth—were reflected in the highly varied commentary on the event, and in the wide range of proposed remedies.[14]

Of course, this filtering mechanism operates not only on expert knowledge, but on virtually any type of information that could potentially come to a policymaker's attention: media stories, communications from external groups or the general public, and, on occasion, even specific events. When discussing a sexual harassment complaint, for example, opponents of such laws and regulations (who tend to believe that complaints are seldom justified) are far more likely to want to know details of the complainant's personal life—social and sexual history, potential motivations, the possibility of financial gain, and so on—than are those who regard harassment as a legitimate social problem.[15] In much the same way, hawks and doves filtered information about the 1968 Tet offensive in Vietnam quite differently. Supporters of America's Vietnam policy were much more likely to focus on the condition of the North Vietnamese and the Viet Cong forces following the attack (generally regarded as a serious military defeat for the North). Opponents of the war, on the other hand, tended to emphasize the disparity between the size and the scope of the attack (as well as the fact that it had occurred at all) with official pronouncements that the enemy was near total defeat.[16]

Participation

The definition of an issue serves another important gate-keeping function by determining which groups, organizations, and individuals—both in and out of government—are to be admitted into the policymaking process. Establishing the Atomic Energy Commission as a civilian agency had the effect of defining the issue of nuclear power far more broadly than the defense establishment had advocated. This, in turn, made it far easier for nonmilitary stakeholders—such as the power industry—to advance their views on federal nuclear R&D policy.[17]

Similarly, when Congress first passed the Federal Insecticide, Fungicide, and Rodenticide Act in 1947, the regulation of such products was almost exclusively a concern of those industries and the agricultural community. By the time that the law had come up for renewal 25 years later, however, the definition of pesticide use had shifted significantly, changing from an agricultural to an environmental issue. One major result of this transformation was that an entirely new set of political actors now became part of the policy debate.[18]

As in the case of problem ownership, there are many instances in which supporters (or, for that matter, opponents) will seek to define—or redefine—an issue in a particular way so as to bring in new participants, or, sometimes, to marginalize current ones. Altering the framework of the policy debate may allow political actors to seek out new sources of support (or opposition).[19] This appears to be what happened, for example, in the case of industrial biotechnology. Many of the earliest public discussions of this issue seem to have been dominated by environmentalists or other opponents, who described the creation of new organisms through genetic manipulation almost exclusively as a (largely negative) ecological matter. Not surprisingly, a significant portion of the public viewed the technology with some degree of trepidation.[20] Over the past decade, however, supporters of biotechnology have succeeded in reframing the political debate, transforming it from an environmental safety issue to a question of industrial development. This, in turn, served to introduce an entirely new set of interests and stakeholders, most notably state and local officials looking for new sources of economic growth and tax revenue. Before long, the technology's environmentalist opponents found themselves completely outnumbered and effectively removed from the policy discussion.[21]

Issue Visibility and Prioritizing

Although there are a few policy areas, such as national defense, that will always be among any government's top priorities, most issues rise and fall on the nation's policy agenda in a seemingly random fashion. At one time or another, pundits and pollsters have labeled health care, abortion, the environment, the federal budget, and, most recently, terrorism as one of this country's "most important problems." Within a few months (or even a few weeks), however, another, completely different "critical issue" will have come along.[22]

It is not at all uncommon for advocates of a particular cause to try to maximize their political support by defining their issue in a way that links it with some ongoing public priority (like defense). President Jimmy Carter's efforts to connect his energy policies to U.S. national security through such phrases as "the moral equivalent of war," and "energy independence" is a clear example of such a tactic.[23] The recent efforts of groups representing persons with disabilities,[24] or gays and lesbians, to have their concerns seen as a logical extension of the civil rights movement of the 1960s can be seen, at least in part, as a similar type of strategy.

Even issues that are not, on the face of it, pivotal concerns can move up on the policy agenda, if not in public prominence, if they can be successfully defined as constituting a "crisis" or an "emergency."[25] It is perhaps for this reason that so many issues are described in such a fashion by their supporters.[26] As the next section will show, however, advocates are seldom able, solely through their own efforts, to convince policymakers, let alone a significant segment of the public, that a given problem has reached the critical stage. More often than not, such a determination follows in the wake of some external event. Clearly, the most dramatic demonstration of this phenomenon has been the significant alteration of the American (and much of the world's) policy agenda in the aftermath of the 9/11 attacks.

Issue Evolution

In the United States, significant change in established public policies—including even outright reversals—is by no means an uncommon event. Sometimes, such as the cancellation of Carter's synthetic fuels program shortly after the Reagan administration took office,[27] such changes are the direct result of some larger political or electoral change. In other cases, however, the causes are far more complex.

In 1987, for example, Congress approved construction of the multibillion dollar Superconducting Super Collider (SSC) project in Texas, a decision that was reaffirmed—albeit with increasing controversy—at budget time during each of the next five years. In 1993, however, Congress reversed itself, voting the SSC only enough funding to cover its shutdown costs. To be sure, part of the decline of the SSC's popularity can be accounted for by some of its former proponents literally changing their minds. It was, after all, seriously over budget (estimated costs had risen from $6 to $8 billion), and its leaders had been accused of mismanagement and misappropriation of funds.[28]

Some political scientists, however, believe that the cancellation of the SSC was not just the result of a simple change of preferences among individual decision makers, but was instead caused by a fundamental shift in the terms of the debate. In other words, between 1987 and 1993 the SSC was redefined, transformed from a purely scientific project into a budget issue. Although the program itself had not changed, after 1990 it was increasingly being judged according to different criteria—its expense—than when it had been approved originally (at which time the discussion dealt almost exclusively with the science it would produce[29]). Not surprisingly, this led to a major realignment of its political support.[30]

Although outright reversals are relatively uncommon, there are numerous cases in which the political dynamics surrounding an ongoing public program have been dramatically altered by issue redefinition. The renewal of the Federal Insecticide, Fungicide, and Rodenticide Act referred to earlier proved to be far more difficult in 1972 than its original passage in 1947. This was primarily because pesticide manufacture and use, seen during the 1940s simply as an agricultural issue (and thus of relatively narrow concern), had been redefined as an environmental problem, which had the effect of opening the law to scrutiny by a completely different set of organizations, institutions, and interest groups.[31]

For a time, the Department of Energy's fusion energy program actually benefited from redefinition. Because it appeared to represent a completely clean and virtually limitless power source, it was essentially "promoted" from a series of (relatively) small research projects in basic physics into a full-scale energy development program during the energy crisis of the 1970s, resulting in a substantial increase in its status, visibility, and, of course, budget. The change proved to be short-lived, however, since a decade later it was redefined yet again, this time into a budget issue—somewhat like the SSC—and subsequently downgraded.[32]

Like the designation of a "crisis," the evolution of a public issue can be triggered by external events, such as natural disasters, actions by a foreign power, or some other unforeseen occurrence. The conversion of the Alaska Pipeline from an environmental to an energy issue, for example, was greatly facilitated by the oil embargo of the early 1970s.[33] Similarly, the accident at the Three Mile Island nuclear power plant in Pennsylvania caused the definition of civilian nuclear reactor policy in the early 1980s to head in the opposite direction, that is, from an energy program to an environmental problem.[34]

Of course, there are also cases in which a redefinition was intentional, the result of efforts on the part of policymakers or other so-called issue entrepreneurs seeking to change the pattern of political support on an issue. The transformation of biotechnology from a "potential environmental hazard" into an "economic development opportunity," apparently due largely to efforts by the biotech industry, represents one such successful effort.[35]

Once again, however, there clearly are limits to an individual's or organization's abilities to bring this about on their own. To begin with, the current "owners" of an issue will almost certainly resist any attempts to redefine it out of their domain, such as when HHS sought to retain Head Start's designation as a welfare program. There are also a number of examples, such as Carter's "moral equivalent of war" appeal for energy conservation and research, where a substantial proportion of policymakers, as well as the public at large, regard the proposed redefinition skeptically.

Multiple Definitions

All too often, governments have been known to pursue initiatives that, particularly when viewed over the long run, turn out to be mutually exclusive. The general tendency among analysts is to regard such "contradictory" policies as a form of bureaucratic or managerial dysfunction, the product either of independent organizations pursuing their own separate policy agendas, or of insufficient attention—if not outright carelessness—on the part of higher-level officials.[36]

It is also possible, however, that the appearance of policy conflict is due to the presence of multiple—and perhaps competing—issue definitions. In cases where the relevant problem definitions have simply been assumed, rather than openly stated—which is usually what happens in government—outside observers and even other policymakers might well regard actions based on a different set of definitions as being either erroneous, incomprehensible, or even corrupt.[37]

In the case of nuclear power, groups that sought tighter regulations and greater controls to enhance public safety for many years felt that opposition from government agencies like the Atomic Energy Commission (AEC) reflected the fact that they represented industry groups. A somewhat more benign explanation, however, might be that the AEC originally saw its mandate as the promotion of civilian nuclear energy, not its regulation. In other words, the way the agency defined its mission was quite different from—if not diametrically opposed to—what the nuclear safety groups expected.[38]

Issue Specificity

Another source of confusion over policymakers' intentions is the fact that traditional policy labels such as "national defense" or "economic development" actually encompass a very wide range of activities. "Welfare" programs, for example, can consist of initiatives intended to alleviate some of the hardships of poverty (e.g., Aid to Families with Dependent Children), as well as "curative" programs such as job training or special education (like Head Start).[39] Thus, simply calling a particular endeavor a "welfare" program may not make it immediately clear which specific aspect of the poverty problem it is intended to address.

Similarly, the term *agricultural policy* refers to government efforts to ensure the stability of particular markets (one of the justifications of price supports), encourage farm ownership (through low-interest loans), promote exports, or facilitate the production and diffusion of agricultural R&D.[40] Thus, while it was suggested above that government programs toward tobacco could be defined as "agricultural" policies, that label by itself could mean a number of different things: subsidies paid to tobacco farmers, say, or perhaps the research efforts to find alternative uses for the crop.[41]

Form Follows Function

Although the stereotypical view of government bureaucracy is that all public agencies look alike,[42] it has long been known that some types of organizational structures are better suited for some types of tasks than others.[43] A highly centralized and formalized arrangement might work well for a military organization, but would almost certainly not be suitable for a scientific research laboratory.[44] This suggests that the appropriate design for a government agency is (or at least ought to be) based on the type of problem it is attempting to solve. The Social Security Administration, for example, has found that it cannot employ the same set of procedures in each of its benefit programs. Decision making in the Old Age Program can often be based on the simple application of a rule, such as whether a claimant is of qualifying age. Such a procedure, however, is not appropriate for the Disability Insurance Division, where decisions on claims represent judgments that are based on an interpretation of medical data.[45]

This connection between form and function becomes especially important for those policy areas possessing multiple problem definitions, or where the "official" definition is subject to change. The policy approach that follows from defining AIDS as a medical problem is very different from one that regards it as, say, a health education issue. Obviously, serious problems can arise when an issue's owner discovers that it is not equipped to deal with a policy's new "identity."

To take only one example, the Department of Energy has faced a major organizational and managerial challenge since the end of the Cold War.[46] Until 1995, approximately half (and, during the Reagan administration, slightly more than half) of DOE's budget was devoted to nuclear weapons production and maintenance.[47] Unfortunately, after more than four decades of such activity, a number of Energy's defense-related facilities, such as Rocky Flats in Colorado and Hanford in Washington State, have developed serious environmental problems. Most of these stemmed from inefficient and short-term methods used to dispose of radioactive and other hazardous wastes. In 1990, DOE established an Office of Environmental Restoration and Waste Management, charged with overseeing cleanup efforts at all of the department's defense sites.[48] As its role in overseeing the U.S. nuclear arsenal became relatively less important after the collapse of the Soviet Union,[49] the effect of this newly emerging concern over environmental restoration had the effect of transforming DOE's image—virtually overnight—from an essential element in American national defense into that of an unusually egregious violator of the country's pollution laws.[50] Adapting to this new mission—not to mention this new image—has placed a great strain on the organization.[51]

Taken together, the points raised above suggest that some of the key questions for understanding how NASA's mission changes over time include:

1. How are government officials defining space policy at different points in time? Do they see it as an R&D issue (to be managed primarily by scientists), a matter of national security and defense (requiring high-level scrutiny and support), or as an economic issue best left to the private sector? In short, what do U.S. policymakers believe to be the nation's space "problem"?

2. How have the patterns of participation around the space program changed over time? Which sets of interests seem to be gaining or losing access to the decision-making process?

3. What sorts of "turf" battles have developed as a result of changing issue definitions? Which individuals or organizations appear to be driving these conflicts? In other words, what effect do external and internal political, economic, social, and other kinds of change have on perceptions of NASA and its mission?

Goals

Unfortunately, even knowing what kind of problem NASA is expected to address does not by itself provide much insight into how it selects specific projects. As the previous chapter noted, most of the criticism of U.S. space policy has been directed at individual programs, such as Project Apollo, the space shuttle, or the International Space Station. Thus, in addition to understanding how space policy has been defined at different points in time, it is also necessary to examine how NASA goes about choosing a "solution." This, in turn, means determining how the agency develops and modifies its long- and short-term goals.

Multiple Goals

Any organization, public or private, always has more goals and objectives than are laid out in their charters or authorizing legislation. At a minimum, these include recruiting qualified personnel (indeed, taxpayers have a right to expect that one of a public agency's "goals" is to hire the best employees that it can find), acquiring resources, maintaining an adequate working environment, improving operating efficiency, and so on. Although they are generally never presented as formal objectives, achieving these so-called operative goals is an essential part of reaching an organization's "official goals."[52]

One of the more common complaints about public bureaucracies is that they too often act as though their only purpose is to achieve their operative goals—increase their size, maximize their budgets, and so on—as opposed to carrying out their intended functions.[53] How often this actually occurs is a matter of some dispute.[54] Nevertheless, the point to be made here is that it is important to take note of all aspects of an organization's operation, not just its stated objectives.

Constraints

It makes no sense to discuss an organization's goals without also taking into account the constraints under which it operates. In fact, Nobel laureate Herbert Simon has argued that meeting some types of constraints really ought to be considered goals in and of themselves.[55] For example, business firms that succeed in keeping their operating costs below a certain limit are both satisfying a constraint and achieving an objective.

As was the case with operative goals, there is a belief that some public bureaus become overly fixated on their constraints. Sociologist Robert Merton wrote of a phenomenon called "objective displacement," whereby an organization's rules, procedures, budgets, and other limits, "displace" its official goals, and become de facto objectives.[56] According to Merton, this is much more likely to occur in organizations where the "real" goals are vague, unclear, or simply difficult to attain, but the constraints are relatively clear and concrete.[57]

Not all constraints are political or budgetary in nature. R&D organizations like NASA face the added problem of physical limits, scientific laws, and so on. DOE's fusion energy program saw its budget and status grow during the 1970s and early 1980s primarily because policymakers believed that the technology could reach the commercialization stage by the end of the century (if not sooner). As it turned out, however, the scientific and technical problems associated with initiating and sustaining a fusion reaction were more difficult than physicists had anticipated.[58] As will be seen, one of the major challenges for science- and technology-based organizations is balancing political demands with scientific and technical constraints.[59]

Finally, there is the problem of "technical lock-in," in which an organization finds itself constrained, not by outside forces, but by the "trajectory" of its own past decisions.[60] Everyone agrees that "innovation," "changing with the times," and "adapting to new circumstances" are positive attributes for any organization (and especially for a private firm). Failure to innovate, "keep up," or "adapt" is usually attributed to such common foibles as "bureaucratic inertia" or personal stubbornness on the part of leadership.[61] It is also sometimes seen as a symptom of organizational decline.[62] According to some economists (and a growing number of business writers[63]), however, some organizations may be unable to make significant changes in their manner of operating simply because they have become so heavily invested in their existing technological infrastructure. Thus, even if they wished to pursue new markets, adopt new technologies, or try new approaches, they may lack the resources to do so: they are "locked in" to the status quo. Ironically, among private companies it appears to be the more successful (which usually means the larger and more established) firms that have the most difficulty adapting.[64]

An excellent example of a locked-in public policy is, once again, the fusion energy program. Most fusion research is carried out on very large machines called "tokamaks" that, when first introduced in the 1970s, seemed to hold out the most promise as a possible commercial device. As a result, dozens of such machines have been built worldwide, and more are in the planning stages. Critics argue, however, that the great expense of building and operating tokamaks has prevented DOE from seriously investigating other technologies.[65]

Individual versus Organizational Goals

Just as an organization can have a number of objectives, stated and unstated, each person associated with it has their own individual goals as well. Hopefully, one of these will be to work toward fulfilling the organization's purpose. Nevertheless, it would naive to ignore the fact that even the most loyal employees have their own personal desires and ambitions with regard to promotion, personal finances, and so on. Obviously, organizations in which the goals conflict with members' personal ambitions may be in for serious problems. Many business leaders have therefore learned that their firms are much more likely to succeed if they are able to structure incentives so that the pursuit of individual goals simultaneously contributes to those of the organization.[66]

For the most part, the literature on the relationship between the goals of organizations and individuals assumes that the latter are organization members. Such a view fails to take into account the possible influence of important external actors, such as (for public agencies) the president, members of Congress, cabinet officers, or other important officials. For example, it is clear that the objectives of an executive branch agency like NASA must conform to the general political goals set by the president. If those goals include reducing federal spending or shrinking the size of government, agency programs will be affected accordingly.

In addition, it does happen from time to time that a public bureau finds itself playing a role in the ambitions of a government official seeking some other political objective(s). These could include an attempt by a member of Congress to gain some logrolling or vote-trading advantage, to acquire favorable publicity or increase his or her chances for reaching higher office. This is certainly not limited to individuals in government. Any external group may see that it is in their interest to support a particular policy perspective. For years, physicists used fears of Soviet science and technology to persuade policymakers to approve funding for ever-larger research facilities, even for fields that had no real Cold War application (see chapter 6).[67]

This last point may sound a bit cynical, suggesting public policy is driven by nothing more than the self-interest of individual public officials or external groups. There is no denying, of course, that political leaders have at times used an issue for their own purposes in a way that was not in the public interest. Perhaps the most (in)famous such case was Senator Joseph McCarthy's use of domestic Communism in the 1950s. Fortunately, there are also numerous examples where the confluence of personal and political objectives is more benign, and even laudatory. Senator Estes Kefauver's investigations into organized crime (which, like McCarthy's, were among the earliest congressional hearings to be televised) were indisputably in the public interest, but also had the effect (almost certainly intentional) of catapulting him into national prominence, and making him, for a time, a leading contender for the 1952 Democratic presidential nomination (he was ultimately nominated for vice president).[68]

Obviously, it is to be hoped that government officials will act on behalf of the public good. Practically speaking, however, it is clear that the motivation to do so will be stronger when such action carries personal benefits as well. From an analytical standpoint, matters become particularly interesting when an issue evolves (see the discussion above) in such a way that these inducements start to diverge.

Thus, along with following the changes in the definition of space policy, it is also important to understand how the space program has been affected by the changing set of goals and objectives surrounding it:

1. How have the goals—personal, political, ideological, and so forth—of participating public officials (as well as external stakeholders) influenced the development of U.S. space policy in general, and of NASA in particular?[69]

2. New presidential administrations—and, on occasion, newly elected Congresses—come into office with new policy goals, as well as a different approach to governing in general. To what extent have these changes affected the course of the U.S. space program?[70]

3. What constraints—political, economic, technical, and so forth—has NASA faced in trying to achieve its objectives?

4. To what extent are NASA decisions "path dependent"? That is, have decisions made at one point in time "locked in," or limited, the agency's future options? What is/are the source(s) of such "lock-ins"? What have officials, both inside and outside the agency, done about them?

Reorganization

Whatever else might be said about it, government reorganization is clearly a popular activity. According to one author, practically every president in the 20th century undertook a major effort to reorganize the administrative machinery of government.[71] Along with these large-scale and highly publicized overhauls—of which the "reinvention" campaign of the Clinton administration and the creation of the Department of Homeland Security under President George W. Bush are only the most recent—have been countless smaller restructurings, wherein new agencies are created or existing ones significantly restructured in some fashion.[72]

Reorganization can take place for a variety of reasons.[73] The most obvious, albeit not the most common, is to provide the means for carrying out some new policy initiative. Presidents Franklin Roosevelt and Lyndon B. Johnson found it necessary to establish a number of new agencies to implement their New Deal and Great Society programs, respectively. Along these same lines, bureaus may also be created in response to the rise of new issues—or new technologies. The Federal Communications Commission, the Food and Drug Administration, the Atomic Energy Commission, and the Department of Homeland Security (not to mention NASA) are only a few among the many organizations established in this century to deal with the growing list of public concerns.

More common is a reorganization intended to promote efficiency or policy effectiveness.[74] This is usually seen in cases where a number of smaller organizations that operate in the same general policy area are combined so as (in theory) to improve policy integration, lower overhead costs, and eliminate duplication and overlap. These were the reasons (at least reportedly) that President Richard M. Nixon established the Environmental Protection Agency and that President Jimmy Carter created the Department of Energy. This same reasoning is sometimes heard in proposals to establish a cabinet-level Department of Science.[75]

Sometimes a chief executive, be it a president, governor, or mayor (or, for that matter, a chief executive office or a corporate director) will "reshuffle" the organizations under them in order to "shake up" what they see as an "entrenched" bureaucracy. Reagan administration officials, for example, often complained that they were unable to carry out their policies because of the "resistance" and "inertia" of the career civil service they had inherited from previous Democratic administrations.[76] One potential solution to such a problem is to break up existing organizations, thereby eliminating any preexisting routines or informal arrangements. Alternatively, a chief executive may attempt to bypass a recalcitrant agency by creating a new organization to carry out his or her programs.[77]

Finally, some reorganizations are carried out for purely symbolic reasons. Elevating the status of an organization by, say, moving it into the cabinet is sometimes seen as a way of demonstrating priorities or establishing the importance of the policy area. Some authors argue, for example, that it was such symbolism that led President Carter to create the Department of Education[78] and President George H. W. Bush to establish the Department of Veteran's Affairs.[79] In addition to symbolism, advocates often claim that the heightened stature will result in the issue receiving more attention and, ultimately, more resources. The evidence for this, however, is mixed at best.[80]

An important point to make about reorganization is that it can be, from the point of view of the affected agencies, exceedingly disruptive. Indeed, one author has likened it to major surgery.[81] It is therefore advisable that it be undertaken only when the likely long-term benefits outweigh the short-run costs (which can include loss of morale and decreases in productivity).

Organizational questions related to NASA's mission might therefore include:

1. Why was NASA created? Was the decision to establish an independent, civilian space organization motivated by policy concerns, a drive for

efficiency, or symbolism? What about subsequent reorganization and reform within the agency?

2. To what extent have efforts to change the course of the space program been accompanied by significant reorganization, either of NASA or elsewhere within the federal bureaucracy?

Conclusions

The three concepts discussed in this chapter—problem definition, goals, and reorganization—provide a great deal of insight into the development of the U.S. space program in general and of NASA's role in particular. The following chapters will show how the early struggle over American space policy (including the basic question of whether the country should even have a space program) grew out of an inability to form a consensus over a definition of space that would justify government involvement. By the late 1950s, after an international event turned space into a major political issue virtually overnight, a definition did emerge—one that was to have a profound effect on the course of the program—helped along in no small part by the personal and political goals of a number of elected officials. This definition, in turn, played a decisive role in shaping the organizational structure of U.S. space policy, particularly with regard to the creation of NASA. All of this was to set the program on a trajectory that would prove very difficult to alter later on: when the definition of space policy was changed (for the most part) during the 1980s, it raised serious questions about the proper role of the nation's space agency.

To lay this argument out fully, however, it is necessary to go back and review some of the major events in space history that predate the establishment of NASA. Conventional accounts of the development of U.S. space policy tend to treat the launch of *Sputnik* in 1957 as a sort of watershed: it is the shock and surprise caused by the Soviet achievement that supposedly led directly to the creation of NASA, the start of the manned space program, and, eventually, Project Apollo. As the next chapter will show, however, matters are somewhat more complicated than this account suggests. To begin with, American reaction to *Sputnik* was, to a large extent, shaped by a unique combination of political factors. More important, many of the decisions made by government officials well before 1957 actually helped set the stage for what was to follow. In short, NASA's "first mission" grew almost directly out of the origin of the Space Age itself.

PART II

FIRST MISSION

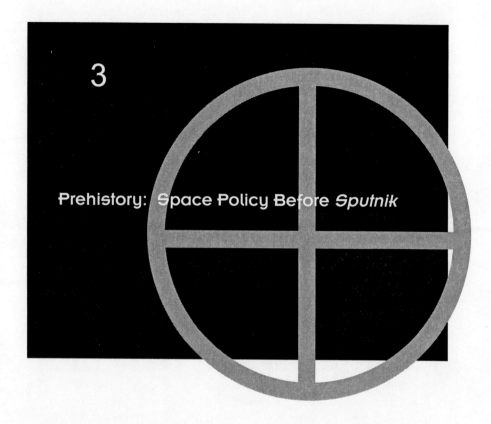

3
Prehistory: Space Policy Before *Sputnik*

Anyone reviewing the historical writing on space exploration might fairly conclude that it is one of those rare human enterprises that has *two* beginnings. Many books and articles trace the origins of modern spaceflight to the theoretical treatises and engineering work of men like Konstantin Tsiolkovsky, Robert Goddard, and Herman Oberth (and later Wernher von Braun). For the most part, these works are part of the much larger literature in the history of science and technology.[1]

There are other authors, however, who see the birth of space travel not just as the development of hardware or the formulation of scientific principles, but rather as the creation and dissemination of an idea. According to this view, the real beginning of spaceflight is to be found in fiction: Edward Everett Hale's "The Brick Moon" (1869), Jules Verne's *From the Earth to the Moon* (1865), H. G. Wells's *First Men in the Moon* (1897), to name a few.[2] These and other novels and short stories played a major role in opening the Space Age, by "inspiring" scientists and engineers in their youth, and later by "popularizing" the notion of space travel.[3]

Despite their obvious differences, both of these paths to space have two important features in common. To begin with, they both convey a certain amount of innocence, if not naivete, particularly when compared with the events to come. Everyone involved—in both fact and fiction—always seem to be acting with the very best of intentions. Space travelers in stories are usually depicted either as scientists engaged in a search for new knowledge, or as ordinary people seeking adventure.

This perspective was carried over into the first science fiction films, such as the French *Le Voyage dans la lune* (1902, loosely based on the Verne novel), the American *Just Imagine* (1930), and the British *Things to Come* (1936, based on the novel by H. G. Wells).[4]

In the real world, men like Tsiolkovsky and Goddard appear to have been motivated primarily (although not exclusively) by the ideals of scientific investigation and discovery. Tsiolkovsky's theoretical work, for example, was both rigorous and thorough (his achievements were particularly impressive in view of the fact that he was largely self-taught). Goddard, a trained physicist, was the first to equip a rocket with instruments for recording data about conditions at high altitudes. This is not to say that they—and others—were not also attracted by the romance of space travel. Goddard decided in his youth to go into rocketry after reading Jules Verne. Tsiolkovsky himself wrote science fiction stories, and to the end of his life maintained a bold vision of humanity's future in space. In a 1935 radio broadcast, a few months before his death, he spoke of "[h]eroes and bold spirits [who] will pioneer [the] first space route: Earth-orbit of Moon, Earth-orbit of Mars, and still further: Moscow-Moon, Kaluga [his home village]-Mars."[5]

The coupling of serious science with a sense of wonder was continued by the various rocket and space societies that began forming in the United States and Europe during the 1920s and 1930s. To be sure, these groups were devoted to serious engineering research: the building and testing of various rocket components, the publication of articles and papers dealing with the theoretical problems of spaceflight. Even so, their names—the American Interplanetary Society (later changed to the American Rocket Society), the British Interplanetary Society, and the German Society for Spaceship Travel—suggest that the members' thinking was rather well ahead of their small model rockets.

The German society (known by its initials VfR) was one of the largest of these organizations, and would prove to be one of the more influential. In addition to its impressive technical achievements—in early 1931, it successfully launched a liquid-fueled rocket to an altitude of 2,000 feet—the VfR provided valuable experience, in public relations as well as engineering, for the young Wernher von Braun. It would also later serve as the core of what would become known as the German rocket team (more on this later).[6]

The second characteristic of early space activity—and, in light of subsequent events, perhaps the more notable—is the almost total absence of any sort of government involvement. Science fiction writers of the 19th and early 20th centuries, even at their most creative, never came close to envisioning anything like NASA, or indeed any other public agency. Usually, these imagined trips into outer space were financed by wealthy individuals or private organizations (such as the Baltimore Gun Club in Verne's *From the Earth to the Moon*), or else were made possible by some fantastic discovery or revolutionary technology (e.g., Wells's "cavorite" in *The First Men in the Moon*), that made external support unnecessary.[7]

Although there was a small amount of public funding for actual rocket research—Goddard received $5,000 from the Smithsonian Institution in 1916, and the National Academy of Sciences provided some support to a rocket development team at the California Institute of Technology (Caltech) beginning in the late 1930s[8]—the initial work in this field remained almost exclusively in private hands. The space societies raised money from a variety of sources, somewhat like today's nonprofit organizations, but, in stark contrast to their descendants half a century later, do not seem to have engaged in any sort of lobbying activities.[9] The major exception here was Hermann Oberth, who actively courted German public and military officials. The most notable such encounter took place in 1929, with a young officer named Walter Dornberger, who would later head up Germany's military rocket program.

Oberth's successes notwithstanding, it is highly unlikely that any government (at least outside Germany) would have seriously considered a request for large-scale funding for rocket and space research during this period. Certainly, public institutions have upon occasion sponsored epic adventures and great expeditions, such as the voyage of Lewis and Clark in the United States. In addition, by the early 1930s, the United States had been supporting scientific research and technological development, most notably in agriculture and medicine, for some time, and had recently begun providing support to civil aviation through the National Advisory Committee on Aeronautics (which would later form the core of NASA).[10] Such ventures, however, had never been undertaken simply for their own sake; rather, officials regarded them as directly contributing to some larger public goal.

It is extremely difficult to see how rocket technology, as it existed at that time, could have met this criteria. Rocketry was very much in the formative stage, and generally unreliable (most of the early test launches, in fact, were failures). Even if it could have been construed as serving any conceivable public purpose, initiating a major program to develop a rocket capable of sending even the smallest object any appreciable distance (let alone a satellite or a human being into space) would have been a perilous, uncertain, and highly expensive proposition for any government. Moreover, any such effort would almost certainly have encountered serious political opposition, particularly from the general public. Goddard's Smithsonian grant was openly ridiculed in the pages of the *New York Times*, and the National Academy of Sciences funding for the Caltech program was reportedly made with great reluctance.[11] Considering also that by 1930, most of the Western world was mired in the Great Depression, public officials would have had an extremely difficult time justifying large expenditures for something like rocketry and space travel.

In short, prior to World War II, governments had little incentive to invest more than a token amount in rocket development. From the point of view of most policymakers, it was—to use a concept presented in chapter 2—an undefined technology, and was therefore, for all practical purposes, invisible. It would not remain so for much longer.

First Definition and the Loss of Innocence

World War II changed forever the status and the nature of rocket research and, ultimately, of space exploration. Almost all the nations participating in that conflict made some use of rocket technology, either as field weapons (really just another form of artillery) or as part of the propulsion system in combat aircraft. Thus, for the first time, governments saw a compelling reason to invest in rocket research and development.

Given the efforts of men like Hermann Oberth (who clearly stands as one of the Space Age's first "issue entrepreneurs"), it is hardly surprising that it was in Germany that the rocket achieved its greatest—and most horrifying—success. In 1932, Walter Dornberger was appointed to head the German military rocket program. A staunch enthusiast for rocket research ever since his meeting with Oberth three years earlier, Dornberger immediately recruited 20-year-old Wernher von Braun as his technical director, and soon brought along much of the rest of the VfR membership as well. It was from their base at Peenemünde, a peninsula near the Baltic Sea, that the German rocket team developed the V-2, the world's first true ballistic missile.[12]

Even though their work involved building weapons of mass destruction for the Nazi regime—and despite the fact that the production of the V-2 was carried out by slave labor[13]—the higher-ranking members of the rocket team always maintained—at least after their immigration to the United States—that their primary interest was actually unchanged from the days of the VfR, that is, developing the technology for space travel. Working for the military, they would later say, was simply a way of acquiring the necessary resources:

> We didn't want to build weapons; we wanted to go into space. Building weapons was a stepping stone. What else was there to do, but join the War Department? Elsewhere there was no money.[14]

Von Braun himself echoed this sentiment during an interview in the 1950s:

> We needed money for our experiments, and since the [German] army was ready to give help, we did not worry overmuch about the consequences in the distant future. Besides, in 1932 the idea of another war was absurd. The Nazis were not then in power. There was no reason for moral scruples over the use to which our researches might be put in the future. We were interested in only one thing—the exploration of space. Our main concern was how to get the most out of the Golden Calf.[15]

While the sincerity (not to mention the morality) of the team members' professed political disinterest has been the subject of much debate and discussion,[16] there is no doubt whatsoever about how the "Golden Calf" defined the program. After the first successful test flight, in October 1942, General Walter Dornberger is said to have declared that "the space age has begun."[17] He also stated, however, at a celebration that evening that "our most urgent task can only be the rapid perfecting of the rocket as a weapon."[18] In early 1943, after viewing a film of the test (see below), Adolf Hitler proclaimed that the V-2 would "turn the tide of the war,"[19] and made its production a top priority.

As it happened, however, the V-2 rocket was not a particularly effective weapon. Although its guidance system was far more advanced than anything that had been developed to that point, a large percentage of missiles never reached their intended targets. They did cause a great deal of destruction in London and other European cities, but, unlike a conventional air strike, the damage was random. In the end, the missile's primary contribution to the German war effort was as an instrument of terror.[20] In fact, the term *contribution* may actually be out of place: some analysts have estimated that the resources invested in the development and production of the V-1 and V-2 were equivalent to constructing 24,000 new fighter planes.[21] Albert Speer would later claim that the rocket program had been "a mistaken investment."[22]

Ironically, Hitler had initially been quite skeptical of the V-2's military potential. Reportedly, his mind was changed after a briefing in early 1943 by none other than Wernher von Braun, who provided "enthusiastic" narration for a color film of the rocket's first successful test flight. Thus, in addition to being the site of the first true ballistic missile, Peenemünde marks the first instance of one other event that—as will be seen in the following chapters—was to become a regular feature of government-funded rocket and space research for the next 40 years: a group of spaceflight "true believers," whose primary—indeed, only—goal was to obtain funding for their vision of humanity in space, had successfully persuaded political leaders (or, in this case, the only political leader who mattered[23]) that supporting their program definitely served the government's own short-run objectives. Or, to use the vocabulary developed in chapter 2, the German government had adopted a definition of the rocket program—a high-priority weapons development project—that allowed von Braun and company to preserve the essential elements of their program.

The desire to carry on their work regardless of who was paying for it (or why) appears to have been a major factor in the rocket team's decision, during the final days of the war, to move as many of their papers and as much of their equipment as possible further west (Peenemünde was directly in the path of the Soviet advance) in order to surrender to the Americans. A U.S. Army intelligence report, based on the initial interrogations of the captured specialists, noted their belief that "this country [the United States] was the one most able to provide the resources required for interplanetary travel."[24]

Relocated to the United States, von Braun and his team began building and testing missiles for the U.S. Army, first at White Sands, and then, in 1950, at the Redstone Arsenal outside of Huntsville, Alabama. Renamed the George C. Marshall Space Flight Center in 1960, this facility was to become one of the key organizations in the U.S. space program.

In view of the fact that a former German military program was now in the hands of the U.S. defense establishment, it is hardly surprising that research into rocketry continued to be defined as a military issue, and that policy ownership remained exclusively with the armed services. Soon the army, navy, and the newly created air force were each attempting to build missiles of their own. Unfortunately, this resulted in a high degree of duplication, and the rivalry between the services frequently got in the way of technical progress. Efforts during the late 1940s to merge missile R&D into a unified program, even to the point of building a single test range, sank in a morass of suspicion and infighting.[25]

An even greater difficulty for the rocket program, however, was official skepticism about its military utility, which, given the prevailing policy definition, meant its utility, period. As noted earlier, the V-2, the most powerful rocket then in existence, had had difficulty hitting targets only a few hundred miles away. To be an effective weapon for the United States, however, a missile would have to be capable of reaching—accurately—far more distant targets. In short, the American military required an intercontinental ballistic missile (ICBM) that had at least a reasonable chance of coming close to the target at which it was aimed.

The technical challenges of such a device were, to say the least, formidable. Since at least part of its flight path would carry it outside the atmosphere, an ICBM would need to be built from materials capable of withstanding the heat from reentry. In addition, its guidance system would have to be far more sophisticated than anything developed thus far. Finally, military planners expected that an ICBM would have to carry a nuclear warhead. The atomic bombs used against Japan had weighed five tons each, well beyond the capability of any missile then under development.[26]

These facts led many prominent scientists and engineers to conclude that an ICBM was technically infeasible. In December 1945, Vannevar Bush, the engineer who had played a major role in the Manhattan Project and who was the principal architect of American postwar science policy, told a Senate committee:

> There has been a great deal said about a 3,000-mile high-angle rocket. In my opinion, such a thing is impossible and will remain impossible for many years. The people who have been writing these things that annoy me have been talking about a 3,000-mile high-angle rocket shot from one continent to another carrying an atomic bomb, and so directed [by a guidance system] as to be a precise weapon which would land on a certain target such as this city [of Washington]. I say technically that I don't think anybody in the world knows how to do such a thing and I feel confident that it will not be done for a very long period of time to come. I think we can leave that out of our thinking. I wish the American people would leave it out of their thinking.[27]

The military rocket program's most serious problem, however, was one that frequently confronts emerging technologies: there simply was no potential mission for such a device that could not already be carried out, far more efficiently and with less cost, with existing hardware. The U.S. bomber fleet, particularly with the development of in-flight refueling and the establishment of overseas air bases, was fully capable of delivering bombs of any size—conventional or nuclear—practically anywhere on earth.

In spite of all of these obstacles, rocket research did move ahead, albeit more slowly than some would have preferred. This was due, in part, to significant improvements in the relevant technologies: more powerful engines, stronger materials, and more sophisticated computing equipment.[28] The other—and perhaps more important—factor was political. In 1952, Americans elected a new president, one who would take a somewhat different view of rocketry and its potential benefits.

A Secret (and a Not-So-Secret) Redefinition

Vannevar Bush's declaration notwithstanding, it had become clear by the early 1950s that missile technology would soon reach the point where it would be possible not only to deliver a payload from one continent to another, but to place a satellite into earth's orbit. Such a capability could potentially be put to a number of uses. As early as 1945, the British author Arthur C. Clarke had discussed the possibility of using orbital platforms as a relay for radio and television signals.[29] A 1954 report to the National Science Foundation by the American Rocket Society described in some detail how satellites could contribute to scientific research, long-range weather forecasting, and improvements in communication and navigation.[30]

Once again, however, it was the needs of the defense community that persuaded the federal government to pursue satellite research. It is useful at this point to recall the discussion of issue specificity from chapter 2. To say that rockets could serve no useful role in "national defense" ignores the fact that this term actually covers a broad range of activities. In addition to the development and procurement of weaponry, it can also refer to the recruitment, training, and management of military personnel, the development of strategies and tactics, and the acquisition of intelligence and reconnaissance data. Thus, the fact that rockets could not serve as an effective weapon (which was becoming less true with each passing year) did not preclude their contributing to U.S. national security in some other capacity.

Like many Americans who had felt the shock and dismay of the Japanese assault on Pearl Harbor, President Dwight D. Eisenhower was keenly aware of the possibility of a sudden, sneak attack on the United States.[31] Such concerns were heightened by a 1955 report from a special panel of the Office of Defense Mobilization. Chaired by James Killian,[32] the main finding of the panel was that while the United States had an "offensive advantage" over the Soviet Union, it was still vulnerable to surprise attack, a situation that could actually lead the Soviets to launch a preemptive military strike. One of the most important ways of preventing this, the report concluded, was through more effective intelligence gathering.[33]

It is therefore hardly surprising that the Eisenhower administration began, through a variety of means, to try to learn as much about Soviet capabilities and intentions as possible. At the 1955 Geneva Summit, the president proposed his famous Open Skies policy, which would permit the United States and the USSR to conduct aerial surveillance of each other's territory.[34] Soviet rejection of the proposal led him to approve secret (and, under international law, illegal) reconnaissance missions over the USSR, first by high-altitude balloons outfitted with automatic cameras (known by the code name GENETRIX), and later by the U-2 spy plane.

Neither of these approaches, however, was fully satisfactory. The balloons, which were entirely dependent on prevailing winds, were largely ineffective: only 44 of the more than 500 launched were ever recovered, and several came down (or were shot down) over Soviet and Chinese territory.[35] The U-2 could, of course, go anywhere a pilot took it, but since this required placing American personnel at risk (not to mention the fact that the flights were a blatant violation of Soviet air space), Eisenhower was never completely comfortable with the plane.

Those who were familiar with the technology, however, believed that satellites could overcome all of these difficulties. As early as 1946, a RAND Corporation study had noted that a "satellite offers an observation aircraft which cannot be brought down by an enemy who has not mastered similar techniques."[36] In a 1952 article for *Collier's* magazine (see discussion in next section), Wernher von Braun wrote that "the telescopic eyes and cameras" of a space station would make it impossible for a country to hide warlike preparations for any length of time.[37] Although he was writing about a manned facility (clearly an infeasible option in the mid-1950s), the principle was basically the same as that of the RAND study. An automated "spy" satellite, taking pictures from earth's orbit, would be far less vulnerable than the observation balloons or the U-2, and would not expose any Americans to danger.[38] Thus, for the first time ever, a space-based technology was able to provide a high-priority[39] government service better than any existing alternative.[40] In March 1955, one month after receiving the Killian report, the Eisenhower administration approved a top-secret "strategic reconnaissance satellite" program, initially designated as the WS-117L.[41]

Defining space R&D as an intelligence project, however, meant that Eisenhower officials would treat it far differently than they would have a basic science program, or even a weapons development project. In fact, although there was no way this could have been known at the time, this new definition of space policy would—in conjunction with other events that were about to unfold—have a profound impact on the entire future course of the U.S. space program.

Science, Secrecy, and Misdirection

One of the more distinctive—and obvious—features of government intelligence programs is the need for strict security. This usually means, at a minimum, restricting access to information, as was done for the Manhattan Project and the U-2 program. Very few officials, even within the Eisenhower administration, had any knowledge that a spy satellite was under development. In fact, with the exception of an off-the-cuff remark by President Johnson in 1967,[42] the United States did not officially acknowledge that it had such a capability until the mid-1970s.[43]

Keeping a space project secret, however, presented special problems. Throughout the 1950s, public interest in spaceflight was being fanned by a number of events. Like his mentor Hermann Oberth 20 years earlier, Wernher von Braun had been spending much of his time since coming to America trying to build up public enthusiasm—and political support—for space exploration. He wrote a series of articles for *Collier's* magazine about rockets, space stations, and human flights to the moon and Mars, all of which managed to ask, in one way or another, "What are we waiting for?" He also collaborated with Walt Disney on three 1-hour television episodes devoted to space exploration, and designed (with science writer Willy Ley) a scale model of an "atomic rocket" for the Disneyland theme park in California. Besides making spaceflight, in political scientist Howard

McCurdy's words, "seem real" to millions of Americans, von Braun was seeking to put pressure on the U.S. government to step up its fledgling space programs.[44]

Meanwhile, far more significant events were unfolding within the scientific community. A few years earlier, the International Council of Scientific Unions had helped set up what would become known as the International Geophysical Year (IGY). Its purpose was to bring together scientists from around the world (more than 60 countries ultimately participated) to conduct an intensive scientific investigation of the earth, including its poles, oceans, deep interior, and—significantly— the region of nearby space.[45] Many scientists and engineers (including von Braun) began lobbying the federal government for permission (as well as funding) to develop and launch a scientific satellite as part of the U.S. contribution to IGY.[46]

The administration soon realized that, given all of this activity, simply keeping silent about satellite research would itself arouse suspicion, particularly on the part of the Soviet Union.[47] Eisenhower officials therefore decided to employ a strategy of misdirection: there would be a second, *public* satellite program, dedicated exclusively to scientific research, that would become the focus of American (and, hopefully, world) attention.

Space historians have known since the mid-1980s (when many of the relevant documents began to be declassified), that this is how the civil component of the U.S. space program began: not in the name of scientific discovery or of conquering a "new frontier," but as an adjunct to—and a cover for—the Eisenhower administration's top priority in space, the spy satellite.[48] Dwayne Day, one of the leading authorities on military space programs, has put the matter quite bluntly:

> Dwight D. Eisenhower, from the first time he gave serious consideration to the concept until the day he left office in January 1961, viewed satellite reconnaissance as a precious commodity that he had to protect with "bodyguards" of cover stories, half-truths, misdirections, and diversions. Much of the American civilian space program, for instance, appears to have been a visible, public means of diverting attention from the security-related programs that Eisenhower valued. Even other military programs themselves shielded his top priority: the reconnaissance satellite program.[49]

As it happened, the IGY proposal suited the administration's needs perfectly: a highly visible, *international* event, conducted solely for the advancement of science. Accordingly, on July 29, 1955, the White House announced that President Eisenhower had approved plans for a series of "small unmanned earth-circling satellites" to be launched during the International Geophysical Year. In view of what is now known about the administration's motives, the language of the official announcement comes off as rather heavy-handed:

> The President expressed personal gratification that the American program will provide scientists from all nations with this important and unique opportunity for the advancement of science.[50]

Using the IGY as a cover had one added—and, for Eisenhower officials, extremely important—benefit. Since no nation had ever placed a satellite into orbit before, the issue of territorial rights in space had never even been discussed, much less resolved through international agreement. As their reaction to the Open Skies proposal (and later to the U-2 overflights) makes clear, the USSR was adamantly opposed to any activity that it regarded as encroaching on its territorial sovereignty. Some U.S. officials feared that if the world's first satellite could even be remotely perceived as representing any sort of military threat, the Soviet Union would seek to assert some form of sovereignty claim over its orbital space. Such a restriction would, to say the least, seriously undermine the reconnaissance satellite program.

If, on the other hand, the first object in space was merely a scientific instrument—better still, if it was part of an international program—such a reaction was much less likely. A satellite launched under the auspices of the International Geophysical Year could therefore help establish an important legal principle that would govern future space activities. A classified 1955 National Security Council (NSC) report, labeled NSC 5520, noted that "[t]he IGY provides an excellent opportunity to mesh a scientific satellite program with the cooperative world-wide geophysical observational program" so as to, among other things, "provide a test for the principle of 'Freedom of Space.'" The report advised that the "U.S. should emphasize the peaceful purposes of the launching of such a satellite," and cautioned that the launch should not "involve any actions which imply a requirement for prior consent by any nation over which the satellite might pass in its orbit."[51]

To be sure, such a satellite would be conducting "real science." NSC 5520, for example, lists a number of legitimate scientific objectives for such a device, such as providing more accurate data on the density and pressure of the upper atmosphere and information on the ion content of the ionosphere.[52] The administration, however, appears to have regarded this as little more than a bonus. The text of the discussion of the "freedom of space" issue in the document far exceeds that devoted to the science involved.[53] It is also highly questionable whether the frugal Eisenhower would have approved the project on the basis of its scientific merits alone, particularly after the costs soared from an estimated $20 million (according to NSC 5520) to nearly $110 million (more than twice the budget of the entire National Science Foundation) by mid-1957.[54] Clearly, this was no ordinary, run-of-the-mill basic science program.

In other words, it was not just the WS-117L that was covered by the intelligence definition of space policy. Although presented to the world as a civilian, scientific enterprise, the IGY satellite was regarded by the White House primarily as a part (albeit an important one) of its space reconnaissance program. This, in turn, meant that (unknown to virtually all of the participants) decisions on the civilian program's development were to be evaluated largely, if not exclusively, according to how they would affect the "main" project, the WS-117L. This fact sheds a great deal of light on Eisenhower's approach to the early U.S. space program, which, as will be seen below and in the next chapter, was soon to come in for some very heavy criticism.[55]

Participation

For years after the United States had "lost" the "race" to launch the first satellite to the Soviet Union—and particularly following the American launch failure a few months later (see the next chapter)—the Eisenhower administration would be roundly condemned for (among other things) not selecting Wernher von Braun to head the IGY program. Subsequent historical discussions would take it for granted that the selection of the Naval Research Laboratory was simply wrong (if not an outright "disaster"), a result of politics and interservice bickering.[56] Even Homer Stewart (from the Jet Propulsion Laboratory), the chair of the committee that made the selection, called it "a real boner."[57] Understanding how the administration defined space policy (such as it was), however, makes it clear that the decision—although certainly open to question—did follow a certain logic.

In the summer of 1955, Assistant Secretary of Defense Donald Quarles established an advisory committee (chaired by Stewart) to recommend which of the many satellite proposals to select for the IGY.[58] There were two major contenders: Project Orbiter, from von Braun's team in Huntsville, which would use a modified version of the army's Redstone rocket for a launch vehicle; and Project Vanguard, from the Naval Research Laboratory (NRL),[59] which would launch with an upgraded Viking sounding rocket.[60]

The NRL proposal boasted a more sophisticated satellite and better tracking facilities, but a number of other factors seemed to favor the Huntsville group. The Viking was still in the experimental stage, far behind the Redstone in development, meaning that von Braun stood a far better chance of launching sooner (in fact, he would later tell the U.S. Senate that he could have had a satellite in orbit as early as 1956, a full year before *Sputnik*[61]). In addition, the army team, which consisted of a number of German scientists and engineers from Peenemünde, had more experience than did the navy group. Finally, thanks to his relentless advocacy of space in the popular media, von Braun himself had become America's most well-known (and most effective) spokesman on behalf of space issues.

It is important to remember, however, that in 1955 the administration's top priority was the spy satellite. The primary purpose of the civilian satellite was not just to conduct scientific research, but rather to convey a benign, nonthreatening image of the U.S. program, and to help establish a principle of "freedom in space." The civil project would therefore have to be run by an organization as far removed from the active military as possible. This (still secret) requirement effectively doomed Project Orbiter. Not only was von Braun working directly for the U.S. Army, but the Redstone rocket he proposed to use as a launch vehicle was also slated to serve as the nation's first ICBM.

The NRL was, of course, a navy facility, and had been from time to time directly involved with the development of military technologies (it had, for example, conducted some of the earliest experiments in nuclear propulsion[62]). Even so, it did conduct a great deal of basic scientific research, and the Viking rocket it was developing had no military application whatsoever. Thus, the administration stood a much better chance of achieving its larger foreign policy goals with Vanguard than with Orbiter. Von Braun (and others) may have been "appalled" at the decision,[63] but it did conform perfectly to the prevailing, if largely unknown, policy definition.

Relevant Information

As with its selection of the NRL, the administration would also later be criticized for the (allegedly) slow pace of the satellite program, and particularly for "ignoring" evidence that the Soviets were on the verge of a launch of their own. For more than four years prior to the launch of *Sputnik I* in 1957, the Eisenhower administration had received repeated warnings about the USSR satellite program. A 1953 report (originally requested by President Harry S Truman) by physicist Aristid V. Grosse was one of the first official documents to sound the alarm:

> If the Soviet Union should accomplish this [orbiting a satellite] ahead of us it would be a serious blow to the technical and engineering prestige of America the world over. It would be used by Soviet propaganda for all it is worth.[64]

The "blow to U.S. prestige" phrase also appears in von Braun's 1955 satellite proposal, although he did not mention the USSR by name (referring instead to "some other nation" that might reach space first).[65]

Similar warnings were even sounded within the Eisenhower administration. The NSC 5520 document actually goes into considerable detail about the consequences of the Soviet Union becoming the first nation in space:

> Considerable prestige and psychological benefits will accrue to the nation which first is successful in launching a satellite. The inference of such a demonstration of advanced technology and its unmistakable relationship to inter-continental ballistic missile technology might have important repercussions on the political determination of free world countries to resist Communist threats, especially if the USSR were to be the first to establish a satellite.[66]

This view was endorsed by a memorandum attached to the NSC report by presidential assistant Nelson Rockefeller. In fact, Rockefeller may well have been—probably unwittingly—the first public official to speak of a "space race":

> [T]he costly consequences of allowing the Russian initiatives to outrun ours through an achievement . . . will symbolize scientific and technological advancement to peoples everywhere. The stake of prestige that is involved makes this a race that we cannot afford to lose.[67]

In July 1957, three months before the Russian launch, CIA director Allen Dulles informed Donald Quarles, the deputy director of Defense, that the president of the Soviet Academy of Sciences had bragged that "literally in the next few months, the earth will get its second satellite." Dulles's memo also presented the intelligence community's assessment that "for psychological and prestige factors, the USSR would endeavor to be first in launching a satellite."[68]

Finally, the Soviet missile and launch facilities in Central Asia had been under surveillance by U-2 spy planes since 1956. These missions had given U.S. officials fairly reliable data about the status of the Russian satellite program (as well its progress toward developing a working ICBM), including a number of photographs of the SS-6 rocket, the booster that would launch the first *Sputnik*.[69] In short, as almost all historians now agree, Eisenhower may have been shocked at the public reaction to *Sputnik* (see the next chapter), but he was not particularly surprised by the launch itself.[70]

As already noted, many critics charged that the administration had "ignored" Soviet space activities. Later assessments of 1950s space policy, which are generally more favorable to the president, are often still at least mildly critical of his "failure to appreciate" the psychological impact of the Soviet achievement, a view that even extends to some senior Eisenhower officials, such as James Killian and Richard Nixon.[71] One author states categorically that the administration's neglect was "almost surely generational."[72]

Viewing the U.S. satellite program in terms of how it was defined by officials at the time, however, helps place these presumed "failings" in a somewhat larger context. Chapter 2 noted that the manner in which a public issue is defined determines which sources and types of information policymakers will regard as relevant. Facts, observations, predictions, and so on that fall outside the accepted definition—even if they are accurate—tend to be ignored. Of course, the line separating what is relevant and irrelevant may not always be perfectly clear. It is certainly possible to imagine cases in which "marginal" information is introduced into officials' deliberations, even acknowledged and discussed, but ultimately plays no substantive role in the final decision.

This seems to have been the case with the "racing" aspects of the U.S. satellite program. The sole purpose of the military portion of the project was to develop and deploy a working surveillance satellite system. The primary goal of the scientific program was to aid the military program by shielding it from public view and by helping establish an international legal principle. The question of which country was "first into space" had very little bearing on any of these objectives, except insofar as the country involved launched the "right"—that is, nonthreatening—kind of satellite. Unfortunately, this attitude was restricted to those few officials who had full information about the U.S. program, and were thus privy to the "official" definition. The majority of Americans were completely unaware of their government's strategy, and would quickly develop a very different view as to what was happening in space.

Conclusion

Assessments of Eisenhower's presidency in general—and his space policies in particular—have changed substantially over the years. As will be seen in more detail in the next chapter, political opponents at the time charged that the president was "disengaged" and "out of touch," or with being more interested in saving money than in defending the country. More recent accounts of the Eisenhower years, however, tend to be somewhat more charitable, particularly with regard to his handling of *Sputnik*.[73] Although, as stated earlier, he is still widely regarded as having been "wrong" not to anticipate the psychological fallout of the Soviet satellite, the president is now generally seen as having been "right" in telling the nation not to panic, and in trusting that America's overall scientific, industrial, and economic establishment was far superior to that of the USSR.[74]

The problem definition perspective, however, presents yet another view of these events. As was also noted earlier, the president's apparent inattention to the psychological aspects of *Sputnik* was a result of the way he and his advisors were conceptualizing the space "problem." Future chapters will show that this would not be the last time such a thing occurred. Nevertheless, and even granting the strategic value of reconnaissance satellites, there is some reason to question the president's selection of this particular definition in the first place. Although he would eventually concede the practical benefits of space in communications and weather prediction (and was quick to extol their virtues when the U.S. program began to produce results in these areas[75]), he apparently did not see them during the 1955–1957 period. Instead, he seemed to have regarded the money devoted to the space programs—equated with spending on scientific research and the military—as a sort of "necessary evil" rather than as an investment in a potentially valuable set of technologies.

Thus, while Eisenhower may deserve credit for "being right about *Sputnik*," he can still be faulted for failing to appreciate the more general benefits of space technology. In other words, it can be said that the president made the mistake of defining the space policy "problem" much too narrowly. It was a mistake that would ultimately prove very costly for him personally, and for the country.

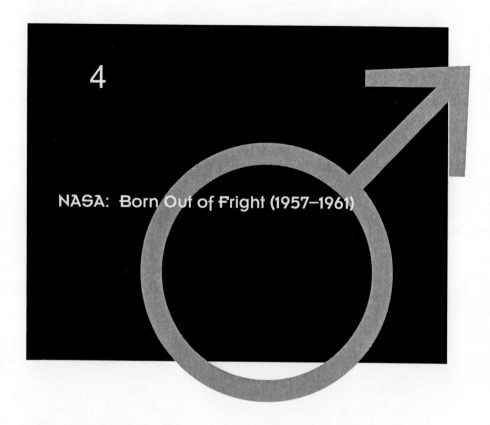

4

NASA: Born Out of Fright (1957–1961)

The twin definitions of space policy into which the Eisenhower administration had settled were utterly demolished by a dramatic series of events in late 1957 and early 1958. As some had warned (see the last chapter), on October 4, 1957, the Soviet Union became the world's first space-faring nation with the successful launch of *Sputnik I* into earth orbit. This historic achievement was followed less than a month later by *Sputnik II*, which carried a living creature—a dog named Laika—into space for the first time. In addition to being space "firsts," these two satellites, weighing 184 and 1,120 pounds, respectively,[1] were far heavier than the 4-pound payload of the navy's struggling Vanguard program. Thus, it appeared to many Americans at the time that the USSR was "ahead" not just on a timetable, but in the development of rocket technology itself,[2] a belief that was only reinforced by the early problems of the (public) U.S. satellite program throughout 1957–1958.

These, however, were not the events—impressive and historic achievements though they were—that served to undermine the Eisenhower space policy. It was, rather, the political activity (or, as some have seen it, the lack of activity) that followed in the wake of the Soviet space exploits that—to use the vocabulary introduced in chapter 2—redefined the American space program. This chapter will show how U.S. reaction to the *Sputnik*s and their descendants brought about a fundamental shift in space policy, and will trace the consequences of that shift, the effects of which are still felt today.

41

Background

Following up on the success of the *Sputnik*s, the Soviet Union continued to stun observers with its even more ambitious Luna series. In January 1958, *Luna I* flew within 3,000 miles of the moon, becoming the first human-made object to orbit the sun. On September 14, 1959, *Luna II* (carrying the Soviet flag) hit the moon 270° from its visible center, and the following month *Luna III* orbited the moon, sending back the first photos of the lunar farside.[3]

Meanwhile, the U.S. program (at least the part that was known to the general public) was literally having trouble getting off the ground. On December 6, 1957, the entire country saw the much-delayed *Vanguard*—its true purpose now superfluous—blow up on live television during a test launch.[4] The first successful American satellite would not come until January 31, 1958, with the launch of *Explorer I*, developed, ironically, by Wernher von Braun and his team in Alabama.[5] By the standards being set by the USSR, however, *Explorer I* was not particularly impressive: it weighed only around 23 pounds, many times smaller than Soviet payloads. In short, even the first U.S. success seemed to underscore how far ahead the Russians were.

Beginning with the announcement of *Sputnik I*, President Eisenhower made a concerted effort to reassure the American public that despite the Soviet Union's apparent lead, the U.S. space program—as well as American science and technology in general—was on par with, if not ahead of, their program. At his first press conference after the launch (on October 9), Eisenhower asserted that the United States and the Soviet Union were not in a "space race," and that Russian achievements in the field did not raise his apprehensions about national security: "not one iota."[6] The following month, he made two nationally televised speeches in the space of a single week on the topic of science and U.S. security,[7] which were intended in part to "put the whole affair in perspective"[8] by reviewing U.S. military strength and presenting "a small sample of our scientists' accomplishments" (such as an experimental nose cone sent "hundreds of miles to outer space [*sic*] and back") to show that American strength "is not static, but is constantly moving forward with technological improvement."[9] He also used these occasions to announce changes like the appointment of a presidential science advisor (see below).

No doubt, the president expected that these statements and policy changes would put the matter to rest. After all, as the former supreme commander of the war against Nazi Germany, he could be counted on to assess correctly U.S. military strength compared to that of any potential adversary. Moreover, as one who had been forced to contend with the destructive power of German rockets, his appraisal of Soviet capabilities in this area would, under most circumstances, have carried a great deal of weight.[10] Finally, although well into his second term, Eisenhower remained highly popular, and had maintained a rapport with the public that, in the words of one biographer, usually came across as "so comforting, so grandfatherly, so calm."[11]

Internal administration documents also make it quite clear that the president felt no particular anxiety about *Sputnik*, nor did he see any need for major changes in U.S. policy. At a meeting with Defense Department officials on October 8 (four days after the launch), Eisenhower noted that "to make a sudden shift in our approach now would be to belie the attitude we have had all along."[12] Similarly, recently declassified minutes of the first National Security Council meeting after *Sputnik*, on October 10, 1957, record that the president himself was anticipating that top aides would soon be obliged to talk to the press, testify before Congress, "and the like," and that under

> the circumstances, he [the president] could imagine nothing more important than that anybody so involved should stand firmly by the existing earth satellite program which was, after all, adopted by the Council after due deliberation as a reasonable program. In short, we should answer inquiries by stating that we have a plan—a good plan—and that we are going to stick with it.[13]

On this particular issue, however, Eisenhower's displays of confidence did not have the desired effect. Instead of reassuring the public, his press conferences and speeches came off as, at best, defensive, and, at worst, in the words of Walter Lippmann, like "a tired old man who had lost touch with the springs of our national vitality."[14] The "not one iota" comment in particular seemed to represent the attitude of one who simply did not understand what all the fuss was about. A Herblock cartoon from early 1958, for example, shows the president hiding from a Russian satellite in bed, with a sign reading "Do Not Disturb One Iota."[15] A 1957 *Life* magazine article, "Arguing the Case for Being Panicky," compared the administration's attempts to calm the public to the boasts of "Persian captains just before Alexander the Great massacred them," or the British army at Trenton during the Revolutionary War.[16] In effect, the president's efforts, far from calming the American people, actually made matters worse: not only did a large segment of the general public still believe that they were facing imminent peril from Soviet rockets, but now many felt that there was a lack of leadership on the issue as well.

Certainly, these efforts were not helped by some top administration officials. Secretary of Defense Charles Wilson (who was in the process of leaving office at the time of *Sputnik I*) once referred to the satellite program as an effort to "get a damn orange up in the air," derided *Sputnik* as a "neat scientific trick," and dismissed any notion of danger from Soviet rockets by saying "nobody is going to drop anything down on you from a satellite while you're asleep, so don't worry about it"[17] Chief of Staff Sherman Adams, who later claimed that he was following the president's desire for "calm poise," ridiculed the notion of an "outer space basketball game."[18] These and other statements only reinforced the view that the administration was made up of elderly men who were out of touch with the new world of rockets and missiles.[19]

Not surprisingly, Democratic officials eagerly (although usually not directly) sought to encourage this view (see below). What is somewhat more unexpected, however, is the extent to which this perception was held by many prominent Republicans as well. In his memoirs, for example, Vice President Richard Nixon expressed some frustration with what he felt was the president's and Chief of Staff Sherman Adams's failure to grasp the importance of space technology.[20] In fact, some authors have gone so far as to suggest that a sort of "generation gap" had emerged within the administration, with the "younger men," such as Nixon and UN ambassador Henry Cabot Lodge, better able to understand *Sputnik*'s significance.[21]

For most historians, the Eisenhower administration's response to the early Soviet space feats represents *the* key turning point in the history of the U.S. space program. According to virtually every study, survey, and historical account—all of which vary widely in their overall assessment of Eisenhower—it was the president's inability to allay the public's anxieties about *Sputnik* that set in motion the whole chain of events that ultimately led up to the establishment of NASA. A 1990 history of the House Science and Astronautics Committee speaks of how "Congress seized the initiative" in setting space policy in 1958 because the executive branch had "sounded an uncertain trumpet." It also notes that "[w]hile Members of Congress were calling for action . . . the public was getting frustrated and infuriated by the 'Papa-Knows-Best' advice" from the administration. The study (which was published by the committee itself, and therefore may be presumed to represent its "official" account) characterizes the Eisenhower administration as "timid" and Congress as "forcefully articulate,"[22] and argues that the president submitted the National Aeronautics and Space Act (the legislation that created NASA) only because he knew that the Congress was about to come forward with proposals of its own.[23]

For their part, those with a less negative view of the administration claim that the NASA proposal represented one of the president's more adroit political moves. According to this view, faced with politically motivated accusations and fear-mongering calls for action, not to mention the largely overblown and irrational reaction to *Sputnik* on the part of the general public, Eisenhower introduced legislation designed to head off what he feared would be "wilder" and more expensive proposals from other policymakers.[24] While this account is somewhat more favorable in its treatment of the president, it nevertheless concludes that the new policies and government reorganization that emerged in late 1957 and early 1958 were the result of a major shift in the U.S. political environment. It is therefore important to understand how and why that shift took place.

The *Sputnik* Difference: External

The impact of *Sputnik* has its roots in a number of "background" factors: the state of the Cold War, the nature of spaceflight, and most Americans' general beliefs about the power of science and technology (at least at that time). Any one of these factors could account for some or part of the public's reaction. As it happens, however, *Sputnik* touched on all of them at the same time.

Some of these concerns have been discussed extensively elsewhere. For example, almost all histories of the period point out that the American public of the 1950s had developed an unusually strong sense of complacency. Although the decade certainly had its share of economic and social problems, including persistent pockets of poverty, continuing racial discrimination, and occasional economic instability, "never had Americans—never had any people—been so generally and spectacularly prosperous."[25] Thus, most of the country took it for granted that, in addition to being the richest, freest, and most powerful people on earth, they were the best educated and most technologically advanced as well.[26] *Sputnik* abruptly called all of these assumptions into question.

Likewise, many authors have noted that a major element in the American reaction to *Sputnik* was the fact that Russia, not their own country, had officially opened the Space Age. The event itself was seen everywhere as an act of enormous historic significance in its own right, the fulfillment of one of humanity's most long-held aspirations.[27] It touched on what Eisenhower science advisor James R. Killian has called "atavistic, subtle emotions about cosmic mysteries," an "instinctive, human response to astronomical phenomena that transcend man's natural ken."[28] At least one author has suggested that these responses bordered on out-and-out superstition, a throwback to a more primitive time.[29] That such a feat had been accomplished—and such deeply rooted feelings inspired—by their primary rival (and that the first, and much smaller, U.S. attempt failed so ignominiously) was a source of humiliation and shame, as well as alarm, for many Americans.

For the Soviet government, the idea of opening a new era in human history fit in very well with their long-standing efforts to portray themselves as an "advanced," "revolutionary" society. Beginning with *Sputnik I*, the Soviet media, public statements by Russian scientists, and the pronouncements of the leadership—particularly Khrushchev—pointed to their space achievements, and especially the fact that they were "ahead" of the United States, as proof of the superiority of the socialist economic and political system. The official announcement of the first *Sputnik*, for example, concluded that "our contemporaries will witness how the freed and conscientious labor of the people of the new socialist society makes the most daring dreams of mankind a reality."[30] A history textbook published in the USSR in 1960 cited *Sputniks I, II*, and *III*, and the Luna series as "convincing proof of the great superiority of socialism over capitalism, the natural result of the development of socialist society."[31] Another writer claimed that the satellites were the hallmark of a new socialist era, much as the steam engine had signaled the "maturity of capitalism."[32]

American officials and commentators were particularly concerned that the image the Russians sought to convey—a progressive, future-oriented society that transformed itself from a relatively backward country into the world's first space power in just 40 years—would prove especially alluring to developing countries. Although there is no evidence that the Soviet space program ever played any role in causing a nation to "go communist," this fear was very real for many Americans, and was frequently raised over the next decade as a justification for an expanded space program.

Soviet advances in space also raised a number of legitimate (if not entirely realistic) strategic concerns. First and foremost among these was the dramatic evidence that the USSR did indeed possess (as it had claimed a few days before the *Sputnik* launch) a rocket powerful enough to serve as an intercontinental ballistic missile. For the first time in its history, the United States was no longer protected by its oceans from a direct military attack. This new sense of vulnerability left many Americans deeply shaken.

There are, however, a number of elements in the U.S. reaction to *Sputnik* that have not been examined in any depth. The first, and perhaps most important, is the unique nature of the Cold War itself. When fighting a conventional (i.e., a "hot") war, a country naturally makes use of all of its resources: industrial capacity, research and development establishments, the talents of its population, and so forth. These assets, however, are usually placed in the service of—and are subordinate to—a single institution, the military. What was different about the Cold War was that virtually every political, economic, social, and cultural institution in the United States was itself seen to be in direct "competition" with its counterpart in the USSR. In other words, it was a "war" that was "fought" on a multitude "battlefields"—political, economic, ideological, cultural—simultaneously:

> What *Sputnik* did, in . . . suggesting Soviet scientific superiority, was to alter the nature of the Cold War. Where it had previously been a military and a political struggle in which the United States need only lend aid and comfort to its allies in the front lines, the Cold War now became total, a competition for the loyalty and trust of all peoples fought out in all areas of social achievement, in which science textbooks and racial harmony were as much tools of foreign policy as missiles and spies.[33]

Eisenhower himself summed up the situation rather neatly in his 1958 State of the Union message:

> [W]hat makes the Soviet threat unique in history is its all-inclusiveness. Every area of human activity is pressed into service as a weapon of expansion. Trade, economic development, military power, arts, science, education, the whole world of ideas—all are harnessed to this same chariot of expansion.[34]

A similar point was made a few months later by Congressman Overton Brooks during floor debate on the National Aeronautics and Space Act:

> The cold war . . . is entering a new phase—or at least a new dimension is being added to it. The major arena in the new, if deceptively peaceful, form of Soviet competition is that of science and technology. . . .
>
> [T]here is in these times a complete blurring between military and civilian consideration. The conflict between nations is a conflict of the total peoples of those nations, not merely of their armies or their capacity for production.[35]

In short, the Cold War was a "conflict" between entire *systems,* with each side seeking to "demonstrate" that some aspect of its educational system, its economy, its artistic establishment—and by implication, its society overall—was superior to that of its opponent. Similarly, an adversary's defect in any of these (or countless other) areas was also flaunted widely because of what it "said" about that country. Both sides considered these portrayals to be especially important for the developing nations of the Third World, which, it was believed, were seeking a "role model" in one of the superpowers.

It is therefore hardly surprising that throughout the Cold War period many different types of actions and events—particularly those that invited a direct comparison between the two countries, or which lent themselves to the ongoing propaganda campaign—often acquired a significance well beyond the intrinsic value of the activity itself. Only under such conditions, for example, could a classical musician—even one as talented as Van Cliburn—be labeled a "hero" in the American popular press,[36] or have his victory in an international music competition (in this case the 1958 Tchaikovsky Competition in Moscow) make front page headlines.[37] Similar sentiments were expressed years later at the 1980 Winter Olympics when, at the height of the Iranian hostage crisis and the early stages of the Soviet invasion of Afghanistan, the U.S. hockey team beat the USSR in the final game. Americans briefly took a greater interest in chess after Bobby Fisher defeated the Soviet champion, Boris Spassky, in 1971 in Reykjavík.

It should also be noted that the importance of these and many such other events has very little, if anything, to do with any practical applications or material benefits. Even the most ardent supporters of Cliburn, Fisher, or the U.S. Olympic team would be hard-pressed to describe exactly how their victories advanced U.S. interests in any concrete or measurable way. Nevertheless, "winning" at these activities seemed very important—indeed, historic—at the time. This view would later infuriate critics of the "moon race" of the 1960s, and while the circumstances may have been very different—no one would have advocated a billion dollar program to train classical pianists—the impulse for winning a largely symbolic race clearly had its roots in Cold War competition of the 1950s.

Although much has been said of the fear inspired by Soviet ICBMs, there has been relatively little discussion of the general attitudes of most Americans toward science and technology in the 1950s. The rapid pace of scientific discoveries and technological advances beginning in the latter half of the 19th century had fostered in most Americans an almost mystical belief in the power of science and technology.[38] Time and again—indeed, within the living memory of many U.S. citizens of 1957—new technologies (developed increasingly, it had seemed, in the United States) had produced drastic changes in the way people lived, worked, and (unfortunately) fought wars. A commonly used plot device in novels, stories, and movies (particularly in works of science fiction) is the hero who overcomes hopeless odds (often at the last possible moment) by making use of some unexpected new weapon, technique, or device.

The *Sputnik*s, combined with the *Vanguard* failure, may have frightened many Americans into believing that the "magic wand" now belonged to the other side: the Soviet Union had access to a revolutionary, exotic, and potentially very powerful new technology that was not yet available to the United States. If most of the public, like the Eisenhower administration, dismissed Khrushchev's claim that *Sputnik* had made the entire American bomber fleet obsolete,[39] the taunt nevertheless may have touched on a deep-seated fear about the ability of new technologies to do just that.

Thus, although the president and his advisors had explicitly rejected the idea of a race with the USSR, many Americans (including a large number of public officials outside the administration) were concerned enough about the apparent Soviet mastery of this new technology to call for the United States to "come from behind" in order to "dominate," "be first in," or "be the leader of" spaceflight. For most of the next decade, it would become difficult to find any discussion of the U.S. space program that did not—at least implicitly—draw some comparison with the Soviet Union. Even President Eisenhower—who would continue to say that such "competition" was "meaningless"[49]—would be forced, when asked directly, to use "racing" metaphors such as in, "our scientists and services have done an extraordinary job in *catching up* as fast as they can."[41] This type of rhetoric would become increasingly common in the years ahead.

The *Sputnik* Difference: Internal (The Power of Definitions)

The preceding chapter noted that even authors who seem generally supportive of the administration's approach to space sometimes fault Eisenhower for "not fully appreciating" the psychological impact of the Soviet satellite,[42] or the propaganda value of "being first" in space. Once again, such claims, at least as they are usually stated, suggest a personal failing on the part of the president. An alternative, and potentially more powerful, explanation can be found by returning to the earlier discussion of problem definitions.

By 1957, the Eisenhower administration not only had a fully developed satellite program underway (albeit behind schedule), but—drawing on the concepts from chapter 2—it seems fair to say that it had a well-articulated definition of space policy in place as well. As the administration saw it, the "problem" in space was designing reconnaissance vehicles, and establishing the proper precedents and interpretations in international law that would permit their use.

Moreover, as also seen in the last chapter, this definition of space policy did not (at least until late 1957) attach any particular importance to being the "first" nation in space. It is not surprising, therefore, that Eisenhower's primary—indeed, only—concern upon hearing of *Sputnik* was its possible military potential, or if it indicated that the USSR was ahead of the United States in developing satellite surveillance technology. At the October 10, 1957, NSC meeting referred to earlier, the only question asked by the president personally was about whether *Sputnik* was capable of taking photographs.[43] It should be noted that this was the only occasion, public or private, in which he expressed any apprehension over Soviet space capabilities (and even then it was not much). Meeting with DOD officials three days earlier, Eisenhower inquired into the status of the air force reconnaissance program, but asked very few questions about the Russian satellite.[44] Indeed, the military intelligence definition of space policy was so deeply entrenched within the administration that air force secretary Donald Quarles actually argued in both meetings that the USSR had "done us a good turn, unintentionally" by establishing the concept of freedom of international space. In short, the administration had conceptualized space policy in such a way that "responding" to *Sputnik*, if it had even occurred to any of the participants at all, would have seemed unnecessary and illogical.

In addition, the fact that an intelligence-gathering program necessarily involves very few people meant that while some administration officials and government scientists were aware of the Russian government's space program, the news of *Sputnik* caught the bulk of government officials (especially members of Congress), and virtually all of the general public, completely by surprise. Having not been privy to (or even aware of) the president's definition, they had been left to interpret these events on their own. As might be expected, given the fears and concerns described above, most chose to define what was for them a new issue as a matter of national security.

This, in turn, had the effect of not only broadening the space program's political base, but also of elevating its political status. To those coming to the issue for the first time, the Eisenhower program—only a portion of which was publicly known—looked paltry and underfunded. The 1955 decision not to assign the satellite project to von Braun's team at Huntsville seemed particularly inexplicable (especially after their *Explorer* beat *Vanguard* into orbit by two months despite the latter's long head start).

These new entrants into U.S. space policymaking, however, actually represent a variety of interests and ambitions. As chapter 2 pointed out, individuals, organizations, and institutions can all pursue a number of goals simultaneously. Thus, it is useful to examine the priorities (besides making the United States "first" in space) and incentives of some of these participants in detail.

Scientific and Technical Interests

The preceding chapter noted that many scientific agencies and institutions had been calling for a greater U.S. commitment to space. Rising public and official concern over *Sputnik* gave these groups another chance to press their case, under circumstances that were, from their point of view, far more favorable. One of the foremost of these advocates was Wernher von Braun. His earlier warnings about national prestige and the possible military advantages to be gained from space now found a new, more attentive audience.[45] Upon hearing about the Soviet launch, von Braun began another major effort to get approval for a satellite launch using the Redstone.

Von Braun was by no means the only engineer making such appeals, nor was rocketry the only field to benefit from *Sputnik*-inspired anxiety. For U.S. scientists, the late 1950s began what a former chairman of the House Science Committee has called the "free ride on the Cold War."[46] National Science Foundation director Alan Waterman, for example, dropped a none-too-subtle hint at a House Appropriations Subcommittee in 1959 when he remarked that "all our scientists who go to Russia come back saying that the government is providing them [Soviet scientists] with the very best facilities."[47] Drawing alarming comparisons with the USSR became commonplace in virtually all scientific fields. The original spur behind the American program to develop nuclear fusion energy was the fear that the Soviets might accomplish it first.[48] Similarly, up until the late 1980s, high-energy physicists generally[49] could count on political support for ever-larger particle accelerators by pointing to Soviet efforts in the field.[50]

Thus, *Sputnik* established the primary political strategy that would be followed by large numbers of American scientists, in some cases literally up until the collapse of the USSR. In one of the pioneering books on U.S. science policy, former *Science* magazine editor Daniel S. Greenberg concluded:

> Waving the red flag to stimulate the appropriations of funds for research and education comprises one of the less admirable parts of the postwar relationship between science and government. The practice can be explained, if not justified, on the grounds that, by and large, Congress and the Executive were more inclined to respond than to innovate, and that a Red scare was the only available device with which science could get the government's attention.[51]

The Democratic Party

Although they had recaptured control of Congress from the Republicans in 1954, the Democratic Party had lost both the 1952 and the 1956 presidential elections by very wide margins (and had just barely held the White House in 1948). Throughout much of the late 1940s and early 1950s, Democrats faced two major problems. First, the GOP had generally been successful in painting the Democrats as "soft" on Communism and "weak" on national defense.[52] Second, the party had a great deal of difficulty keeping its various factions united, particularly as conflicts over racial segregation intensified during the mid- to late 1950s. Not surprisingly, then, they had been unable to make a dent in the president's popularity, which in some polls had been running at nearly 80 percent.[53]

Sputnik seemed to provide a golden opportunity. For the first time in nearly a decade, Democrats could now charge Republicans with failing to meet the Communist challenge. As a new issue that transcended factional divisions, it could serve as a sort of unifying force. It gave the party a way to criticize Eisenhower that did not look petty or partisan, particularly after the *Vanguard* launch failure (as suggested earlier, the Democrats were unwittingly aided in this effort by the statements of some of the president's own advisors). Finally, because it turned space into a major security issue, it provided the means for the Democratic-controlled Congress, where most of the party's most prominent members served, to become involved in criticizing—and setting—policy.

U.S. Congress

As a general rule, representatives and senators do not concern themselves with the day-to-day operations of basic research projects, and most military reconnaissance activity is conducted in near, if not total, secrecy. Thus, neither of the administration's initial, pre-*Sputnik* definitions of space policy lent themselves readily to any sort of congressional involvement. On the other hand, a program seen as having implications for the country's very survival would naturally be of great concern to all members of Congress. It is therefore hardly surprising (particularly when combined with the political opportunities described above) that redefining the space program as a matter of immediate national urgency would lead to a greatly expanded role for the legislative branch.

Both houses of Congress lost little time in creating new committees devoted to space and related activities. The House Committee on Science and Astronautics was established on July 21, 1958, and the Senate Committee on Aeronautical and Space Sciences was created on July 24, 1958 (it was abolished in a Senate committee reorganization in 1977, and its functions transferred to the Committee on Commerce, Science, and Transportation).[54] The House committee in particular developed a reputation for being especially aggressive on space matters, and would at times prove to be something of a headache for the president and NASA.

Lyndon Baines Johnson

These last two institutions, the Democratic Party and the U.S. Congress, come together in one politician, Senate Majority Leader Lyndon Baines Johnson. As a politician who had first come to Washington during the New Deal, Johnson would not have been as concerned as was Eisenhower about government growth, particularly when it came to dealing with a major public problem. That Johnson saw the *Sputnik*s as such a problem (if not an opportunity—see below) there is no doubt. In his memoirs, he wrote of *Sputnik I* and *Sputnik II* producing feelings of shock, apprehension, bewilderment, and a sense of "frustration, bordering on desperation." The fears are familiar: the Soviets might be "ahead" technologically, American prestige had taken a severe blow, Third World nations might begin to see the USSR as the "wave of the future."[55] Johnson wasted no time in acting on these feelings. On the evening of October 4, 1957, he called a number his Senate colleagues, as well as the staff of the Subcommittee on Preparedness (of which he was chair) from his ranch in Texas.[56]

Clearly, it was not lost on the senator that much was to be gained—both for the Democratic Party and for himself personally—from keeping the *Sputnik*s in the public eye. A memo on political strategy from longtime Johnson aide George Reedy, dated October 17 (i.e., less than two weeks after *Sputnik I*), noted that the Russian satellite was the perfect issue for the Democrats. It called the country's current leadership into question, it involved "not only national defense, but also a whole range of scientific achievements (including the possible conquest of the universe itself)," and, best of all, it could be handled in a "completely and sincerely non-political" fashion: "*[i]f the issue has merit, the politics will take care of themselves.*" For Johnson personally, Reedy noted in a cover letter that the issue, "if properly handled," would not only "blast the Republicans out of the water, [and] unify the Democratic Party," but would also "elect you President."[57]

The Reedy memo, especially the "elect you President" phrase, has been cited by some authors as evidence that Johnson's (and, by implication, that all

Democrats') efforts on behalf of the space program were largely, if not exclusively, motivated by partisan political interest.[58] That particular document, however, is not in and of itself conclusive. After all, the concerns expressed were not very far out of line with those of the public at large. Moreover, most participants were making a strong effort at least to appear nonpolitical. The memo itself states that "any effort to inject politics into such an inquiry [of the missile and satellite program] would merely lead to a 'you're another' type of shouting match."[59] In addition, in a meeting on October 9, President Eisenhower warned Johnson that if the Democrats tried to turn the space program into a partisan matter, he would blame the Truman administration for failing to commit any funds to the U.S. missile program before 1950 (and only small amounts after that).[60]

The fact that the *Sputnik*s and the *Vanguard* failure had both occurred on Eisenhower's watch probably made any effort to place blame on an administration that had left office six years earlier a hollow one. Nevertheless, Johnson and others went to great lengths to stress that their concerns were nonpartisan. During his opening statement at the Preparedness Subcommittee Hearings on November 25, he noted that

> Our goal is to find out what is to be done. We will not reach that goal by wandering up any blind alleys of partisanship. . . . There were no Republicans or Democrats in this country on the day after Pearl Harbor. . . . There were just Americans anxious to roll up their sleeves, to close ranks, and to wade into the enemy.[61]

Of course, from the Democrats' point of view, the beauty of the *Sputnik* issue was that it was not necessary to raise politics or partisanship: simply keeping the issue before the public—and keeping that public in a jittery mood—was enough. In the conclusion to his memo, for example, after calling once again for an inquiry "definitely without partisanship," Reedy notes:

> There is, of course, a partisan advantage to turning the attention of the American people to such an issue. *But it is an advantage that would accrue to the American people as well. This may be one of those moments in history when good politics and statesmanship are as close to each other as a hand in a glove.*[62]

This is by no means to suggest, however, that Johnson had no political or personal aims. It is clear—from evidence in addition to the Reedy memo—that he and his staff recognized early on the potential of *Sputnik* to unify the often-contentious Democratic Party (and do so in a way that benefited the senator personally). A memorandum from another staffer, James Rowe, dated November 15, 1957 (less than two weeks before the Preparedness Subcommittee Hearings), urged Johnson to:

> telephone a fairly large number of people in the Democratic Party asking for their views on what you should do. I do not think that it

really matters very much what they tell you—although it is always possible someone might come up with a good idea—but I do think they will be pleased and cooperative if you consult them beforehand.

I would certainly suggest Harry Truman, Adlai Stevenson, Averell Harriman, and others. Personally, I would take the Democratic Advisory Committee list and go right through it, even including some of the more obscure national committeemen.[63]

With regard to Johnson's own ambitions, it did at times appear as though he was racing against his Democratic colleagues as much as the Russians. For example, in preparation for the Senate hearings on the National Aeronautics and Space Act (see below), scheduled to begin on April 16, 1958, Johnson aides became very concerned when House Majority Leader John W. McCormack announced plans to begin *his* hearings one day earlier, on April 15, with von Braun as his first witness. "The implications of this are clear," wrote the aide. "McCormick [sic] is seizing the ball and is going to run with it. I should think that the Senator would not be overjoyed by this."[64]

Perhaps even more telling is a speech delivered to the Democratic Conference on January 7, 1958, in which the senator made a few rather odd claims about what the Soviet lead in space might mean:

Control of space means control of the world, far more certainly, far more totally than any control that has ever or could ever be achieved by weapons, or by troops of occupation.

From space, the masters of infinity would have the power to control the earth's weather, to cause drouth [sic] and flood, to change the tides and raise the levels of the sea, to divert the gulf stream and change temperate climates to frigid.[65]

To make these types of claims immediately after *Sputnik* would have been one thing; as noted earlier, general knowledge about rocketry and astronautics was relatively low at the time. This speech, however, took place after nearly two months of hearings, with, in the senator's words, "34 witnesses, 3,000 pages of transcript, 150 to 200 staff interviews with individuals concerned with our missile and satellite programs, and searching questionnaires sent to industrial organizations, leading scientists and engineers, and leading educators."[66]

The only possible explanation for such hyperbole is the fact that Johnson's personal ambitions required a certain measure of public apprehension. As future chapters will show, this would not be the last time that he would attempt to use the space program in this fashion.

The Founding of NASA

As chapter 2 suggested, a major political upheaval or an important new policy initiative can spark a move for government reorganization. As this discussion has suggested, the events surrounding the *Sputnik*s provided both. Following in the wake of the Soviet satellites were many, many proposals for major administrative and organizational change, arguably the most extensive ever to take place in the United States during peacetime. Obviously, the creation of NASA was one of the more important and far-reaching of these.[67] To appreciate fully the atmosphere into which the first United States space agency was born, however, as well as to understand better the nature of its original mission, it is necessary to look at the whole range of *Sputnik*-inspired reorganization:

- Advocates had been trying for years to obtain federal funding for education, particularly in the sciences. Only *Sputnik* was able to overcome the reluctance of some members of Congress to move in this direction. The National Defense Education Act of 1958 brought the federal government into the public education forum to a far greater degree than ever before.[68]

- In the executive branch, President Eisenhower established the President's Science Advisory Council (PSAC). An organization of this type had actually been created some years before, but was attached to the military. Eisenhower's move served to place scientific expertise—and scientists' interests—directly within the White House itself. He appointed James Killian as the first presidential science advisor.[69]

- On February 8, the secretary of defense, acting on instructions from Eisenhower, announced the creation within DOD of the Advanced Research Projects Agency (ARPA) to direct its rocket and space programs. It would later be renamed the Defense Advanced Research Projects Agency (DARPA) and acquire responsibility for all long-range defense-related R&D.[70]

- Federal spending on research and development increased markedly in the wake of *Sputnik*. At the National Science Foundation, for example, outlays for research and related activities more than doubled, rising from $44.4 million in FY 1956 to $102.1 million in FY 1958.[71] Between 1955 and 1960, total federal spending on R&D rose from $2.6 to $7.4 billion for nondefense agencies, and from $9.6 to $23.0 billion for DOD.[72]

- As noted earlier, both houses of Congress set up new committees devoted to space. For the House of Representatives, this was the first new standing committee created since the Legislative Reorganization Act of 1946 reduced the number of committees from 48 to 19, and the first committee devoted to a new subject since 1892.[73]

- Many other changes were proposed, but not approved. These included a joint congressional committee along the lines of the Joint Committee on Atomic Energy,[74] a cabinet-level Department of Science,[75] and even a Department of Space. Some policymakers called for Eisenhower to appoint a "missile czar" to coordinate all U.S. space and missile development activities.[76]

NASA's basic structure of the agency was established by a PSAC panel convened on Eisenhower's order in February 1958. This group considered a variety of organizational strategies—and listened to requests and demands from myriad competing interests—for managing the nation's space policy. These included turning the program over to either ARPA, the Atomic Energy Commission, a totally new agency, or a reconstituted National Advisory Council on Aeronautics. In March, the PSAC panel endorsed this last alternative, and on April 4, 1958, the president sent the National Aeronautics and Space Act to Congress. After much discussion, debate, and amendment (see below), it was passed by both houses on July 8 and signed by Eisenhower on July 29. NASA officially went into operation on October 1, 1958.

Nevertheless, in view of all of the reorganization described above, establishing NASA really ought to be seen as just one more element—albeit a central one—of a much larger, ongoing process. The level and intensity of the activity in the months following the *Sputnik*s and *Vanguard* resemble nothing so much as a general mobilization. An exceedingly large share of American material and intellectual resources were now engaged in assuring that the country would remain (or, as some saw it, become once again) "first" in education, science and technology, and (at the time, most important) space.

That this was to be the paramount NASA mission is made abundantly clear by congressional debate of the Space Act itself. True to form, President Eisenhower's letter of transmittal mentions the Soviets only in passing, with the simple observation that both countries have "placed in orbit a number of earth satellites," and even that statement is weakened in the next sentence, where he notes that "it is now within the means of any technologically advanced nation to embark upon . . . programs for exploring outer space." The only other reference to international competition is the "preservation of the role of the United States as a leader" clause from the act as one of the responsibilities of the proposed space agency.[77]

Upon reaching Congress, however, the terms of debate shifted radically. In his opening statement as the House Select Committee on Astronautics and Space Exploration began hearings on the Space Act on April 15, 1958, Chairman and House Majority Leader (and future Speaker of the House) John W. McCormack declared:

> The immediate problem is obvious. We see another nation of great potentiality, militant and competitive, which has already made the first advances in the mastery of outer space. We cannot stand by and watch this nation make that mastery complete.[78]

On the Senate side, Lyndon Johnson, chairing the Select Committee on Space and Astronautics, set (as always) an even more disturbing tone:

> What we do now may very well decide, in a large sense, what our nation is to be 20 years and 50 years and 100 years from now and, of no lesser importance, our decisions today can have the greatest influence

upon whether the world moves toward a millennium of peace or plunges recklessly toward Armageddon.[79]

When the Space Act reached the floor of the Congress in early June, such rhetoric continued. In his statement introducing the bill, McCormack began by noting that "[i]t is amazing what can be done in what is called outer space and what they know can be done in 5, 10, 15, or 20 years from now." But in his next sentence he used the language of the Cold War (albeit in a convoluted way):

> Involved in that might be the very survival of our Nation. For example, another country that was a potential enemy of ours or an enemy of a government of laws and not of men, in other words an enemy of the free world, a potential enemy, if it was able to get a decided advantage, that advantage might result in the destruction of the entire world or in the subjugation of the entire world to that particular nation.[80]

Although he did describe some of the other specific benefits of space research—the creation of new industries and employment opportunities, new technologies in weather forecasting, transportation, communications, and so on[81]—he quickly returned to the point:

> The urgency of this legislation can hardly be exaggerated. It is highlighted by the series of surprising developments the Soviet Union has already accomplished in astronautics. The United States must leapfrog these Soviet accomplishments.[82]

This pattern would become familiar throughout the debate over the creation of NASA: proponents arguing that the primary issue was national security (if not "national survival"), but that there were "also" other advantages to be gained from developing American space capabilities. Gordon Mcdonough of California, for example, in noting that NASA's initial budget was to be approximately $100 million, stated:

> These costs . . . are not small sums, and we must be fully aware that we are entering into a program—and an international competition—of great magnitude. . . . The costs are fully justified in any event, for reasons of national survival. But, in addition, there will unquestionably flow from this effort inestimable economic benefits. Many of these cannot now be foreseen, any more than Columbus or his contemporaries could have, in 1492, estimated the world benefits that would flow from his discovery of America.[83]

Similarly, Lee Metcalf of Montana stated:

> My discussion of the race with the Soviet Union to perfect space exploration devices has been couched essentially in terms of national

survival. This is a factor which the Members should not overlook when they come to authorize this new agency and later to approve appropriations for it.

I do not want to end my remarks on this note, however. It is entirely consistent for me to report to you that the real hopes of the space age lie in the field of peace and human betterment.[84]

NASA's paramount role as a defender of national security was further underscored by Senate modifications to the Space Act. Lyndon Johnson proposed an amendment creating, within the executive office of the president, a National Aeronautics and Space Council that would "coordinate" all of the nation's space activities, military and civilian, at the highest level of government.[85] Members of the council were to include the chairman of the Atomic Energy Commission and the secretaries of state and defense, as well as the NASA administrator.

As might be expected, President Eisenhower continued to resist this approach to space policy. He opposed the Space Council proposal, for example, because he thought (significantly) that it would grow into another National Security Council,[86] and that space would "never be that important."[87] These objections held up final approval of the Space Act until he and Johnson were able to work out a compromise whereby the president himself would chair the council, thereby keeping it under his direction and control.

In addition, even as members of Congress were speaking of "racing," "catching up with," and "beating" the USSR, Eisenhower continued to assert that no such "competition" existed. His statement upon signing the Space Act made no mention of the Soviets, and with only a vague reference to "further equipping the United States for leadership in the space age."[88] During the meeting at which he was offered the job of becoming the first NASA administrator, T. Keith Glennan recalled that the president did not refer to Russian space activity at all, but simply said that he wanted a "sensible, well-paced" space program.[89] Indeed, to the end of his term, he maintained his stance against the space race, and would later express his strong disapproval of the expansion of the program conducted by his successors.[90]

By 1958, however, the new definition of space policy had been firmly established, and, with the exception of Eisenhower himself, administration officials had little choice—particularly when dealing with Congress or attempting to reassure the public—but to accept it. Although Glennan would confide in his private diary that "we are not going to attempt to compete with the Russians on a shot-for-shot basis in attempts to achieve space spectaculars,"[91] in public settings (especially congressional hearings) he often made use of the more "fashionable" rhetoric.

Appearing at authorization hearings before the Science and Astronautics Committee in 1959, for example, Glennan gave a "many-sided answer" to the question of "how we justify the expenditure of millions of dollars to explore the unknown." He listed (in order) expanding scientific knowledge, applications, the creation of new industries, and understanding the unknown. "But," he then said, "there is still another and *overriding* reason for our program of space exploration":

I believe it is becoming increasingly obvious to the world that Russia's space activities are devoted, as are most of their activities as a nation, in large part to the furthering of communism's unswerving designs upon mankind. . . .

[W]e know that the Soviet Union is in this business and that their successes thus far have been impressive. *I cannot believe that they will withdraw from a race in which they hold even a slight lead.* Have we any choice, as leader of the free world, but to press forward with diligence on a well-planned program for the exploration of this new environment?[92]

In much the same vein, the *Introduction to Outer Space*, a document produced by the same PSAC panel that recommended the creation of NASA, and intended to aquatint the general public with some of the basic principles (as well as the opportunities) of space exploration, listed the primary justifications of the program as (in order):

1. The compelling urge of man to explore and discover.

2. National defense.

3. Enhance the prestige of the United States among people of the world and create added confidence in our scientific, technological, industrial, and military strength.

4. Scientific observation.[93]

Interestingly, the possible economic benefits of space were not mentioned.

Along these same lines, the first post-*Sputnik* statement on U.S. space policy, written in August 1958 (and originally classified) opens with the observation:

The USSR has surpassed the United States and the Free World in scientific and technological accomplishment in outer space, which have captured the imagination and admiration of the world. The USSR, if it maintains its present superiority in the exploration of outer space, will be able to use that superiority as a means of undermining the prestige and leadership of the United States and of threatening U.S. security.[94]

It lists U.S. objectives in space as being:

Development and exploitation of U.S. outer space capabilities as needed to achieve U.S. scientific, military and political purposes, and to establish the U.S. as a recognized leader in this field.[95]

What these "scientific, military and political purposes" might entail is stated explicitly in an earlier draft of the policy statement:

> A degree of competence and a level of achievement in outer space basic and applied research and exploration which is at least on a par with that of any other nation. . . .
>
> Applications of outer space technology, research and exploration to achieve a military capability in outer space sufficient to assure over-all [sic] superiority of U.S. offensive and defensive systems relative to those of the USSR. . . .
>
> World recognition of the United States as, at the least, the equal of any other nation in over-all [sic] outer space activity and as the leading advocate of the peaceful exploration of outer space.[96]

In short, the transformation of space policy was now complete. The president's misgivings notwithstanding, NASA's mission—to take charge of the Cold War in space—was firmly in place. The next step would be to develop specific programs, the "weapons" with which that "war" was to be fought.

The Mission Made Flesh (Literally)

How this new definition of space policy was to work in practice can be seen most clearly in the first major post-*Sputnik* program, Project Mercury. To be sure, the stated goals of the man-in-space program[97] did contain some legitimate scientific objectives, such as investigating "man's performance capabilities and ability to survive in a true space environment."[98] Nevertheless, there is an abundance of evidence attesting to the true purpose of Project Mercury: having "lost" the race with the USSR to orbit the first satellite, most participants in space policymaking (old and new) had now set their sights on "beating" the Soviets to the next "milestone" in space competition.

Beginning in early 1958, the air force, army, navy, and the National Advisory Council on Aeronautics had each prepared their man-in-space proposals, almost all of which had in common the fact that they could (in theory) be carried out in a relatively short amount of time. Indeed, most proposals seemed quite unapologetic about it (the air force even went so far as to designate one of their proposals "Man in Space Soonest"[99]). Each relied heavily on off-the-shelf technologies, required very little in the way of new technological development, and were to fly on the only large (relatively speaking) U.S. booster rocket available, the Redstone. Glennan's description of Project Mercury was that it would "extend the state of the art as little as necessary."[100]

The justification for such an approach—which was given a "highest national priority" procurement rating (a so-called DX rating) in late 1958[101]—was stated explicitly on a number of occasions, by a variety of officials. The Senate Aeronautical and Space Sciences Committee report declares:

> Meaningful appraisal of this Nation's man-in-space program must invariably be done in context with similar efforts underway in the USSR. The psychological impact of a Soviet "first" in this area could have tremendous effect on world opinion and play an important role in the "cold war."[102]

The same theme was sounded in the report of an ad-hoc panel on "man in space" assembled by Science Advisor George Kisatowsky (who replaced Killian in 1959) at Eisenhower's direction. Although it notes the scientific benefits of space exploration, as well the "challenge" involved, it ultimately claims:

> At present *the most compelling reason for our effort* has been the international political situation which demands that we demonstrate our technological capabilities if we are to maintain our position of leadership.[103]

Finally, in the Eisenhower administration's last official statement on space, originally circulated with a National Security Council designation, but formally released as a National Aeronautics and Space Council document (which, incidentally, contains a lengthy appendix on USSR space program):

> To the layman, manned flight and exploration will represent the true conquest of outer space. No unmanned experiment can substitute for manned exploration in its psychological effect on the peoples of the world.[104]

In other words, most officials—including some within the administration—believed that if the United States were to succeed in placing a man in space first, it would, in effect, negate the accomplishments of the Soviet *Sputnik*s, since these did not represent the "true conquest of space."

There was, of course, some opposition to this use of the nation's space resources. Prominent scientists like Vannevar Bush (who, it will be recalled from the last chapter, had argued against the early U.S. missile program) and even Killian and Kisatowsky, who were in accord with Eisenhower's limited support for Mercury,[105] warned against placing too much emphasis on man-in-space.[106] This period also marks the beginning of the long-standing and deep disagreement between supporters of manned space flight and the space science community (see chapter 1).[107] The antagonism would soon grow even stronger in the years ahead. In fact, they would lead Jerome Wiesner, who would become President Kennedy's science advisor to recommend in 1961 (in a report prepared for the presidential transition) that Project Mercury be de-emphasized.[108]

Such misgivings, however, could not diminish the very high visibility of Project Mercury, nor the Congress' and the general public's perception of its importance. NASA's announcement of the selection of the first Project Mercury astronauts on April 8, 1959, turned seven test pilots into celebrities and national heroes literally overnight.[109] Clearly, the claims concerning manned space flight's "psychological impact" were proving to be correct, a fact that would later come back to haunt U.S. space planners.

A Change in the Wind?

Although American interest in—and anxieties about—space remained relatively high, it appears that the political forces driving U.S. space policy expansion were beginning to ebb somewhat by the end of the 1950s. To begin with, the United States began to make its own inroads into space. Between 1958 and 1961, it successfully placed 30 spacecraft into earth orbit (albeit with 29 failures) and had sent two beyond into interplanetary space (with 10 failures).[110] While it could not yet match Soviet launching capacity in terms of thrust or payload weight, it could claim a number of solid achievements in space science and applications. *Explorer I*, the first U.S. satellite, discovered the Van Allen radiation belts surrounding the earth. Other such successes included the *Echo I* communications satellite and the *Tiros* weather satellite, both launched in 1960. Indeed, some observers went so far as to claim that, in terms of actual results and material returns (i.e., matters that "really counted"), the United States was actually "ahead" of the USSR.[111]

Moreover, by the end of the decade the launching of satellites had become a familiar enough event (although by no means routine) that some of the deeper fears associated with the *Sputnik*s had begun to decline. Indeed, it has been suggested that the "diminishing returns" from so-called space spectaculars caused Premier Khrushchev to push Soviet space technicians too hard, resulting in the launch disaster described above.[112] In short, some of the more "primeval" feelings associated with spaceflight inspired when it was brand new were now wearing off.

Indeed, the tensions had receded to the point where Eisenhower was able, in his final budget submission in January 1961, to recommend a NASA appropriation of $190 million *less* than the agency had requested. Even more surprising was that most of the decrease was to come out of the budget for manned space flight, including Project Mercury.[113] A substantial portion of this funding, however, was restored in March by the new Kennedy administration (although, as the next chapter will discuss, there was to be some controversy over the manner in which this was done).

This easing of American fears almost certainly accounts for the fact that space played almost no role in the 1960 election.[114] To be sure, the status of the U.S. program was mentioned from time to time. The Democratic Party platform, for example, noted that the Republicans had allowed the Russians to "forge ahead" in space.[115] Likewise, the Kennedy campaign's position paper on space stated that "the [Eisenhower] Administration's initial attempts to go into space on a budgetary shoestring have made it difficult to compete,"[116] and Kennedy himself would note in his campaign speeches that "our position in outer space compared to the communist position [is not] as strong as it was some years ago."[117] These statements, however, appear to have been simply part of the overall Democratic strategy to portray Eisenhower as old, tired, and "out of touch," rather than to use spaceflight itself as an issue.

For the first few months of his term, it appeared that President Kennedy largely shared his predecessor's view that space should not be guided by U.S.–Soviet rivalry. Their specific approaches in this regard, however, differed considerably. Whereas Eisenhower simply wanted American space policy to proceed on its own course without regard to what the USSR chose to do, the new president actively sought to promote superpower cooperation in space. His Inaugural Address, for example, called for the United States and the USSR to "explore the stars together,"[118] and his first State of the Union Message, challenged the two countries to work jointly toward trying to "evoke the wonders of science instead of its terrors."[119]

Internal documents strongly suggest that the new president fully intended to put these sentiments into practice. He directed Science Advisor Jerome Wiesner to convene a group of specialists to develop concrete proposals for joint space projects. A draft report from that panel, dated April 4, 1961, stated explicitly:

> The objectives are to confirm concretely the US preference for a cooperative rather than a competitive approach to space exploration, to contribute to reduction of cold war tensions by demonstrating the possibility of cooperative enterprise between the US and the USSR in a field of major public concern, and to achieve the substantive advantages of cooperation that in major projects would impose more of a strain on economic and manpower resources if carried out unilaterally.[120]

Whether or not these efforts represented a sincere desire to collaborate—some authors have suggested that they were simply an effort on Kennedy's part to appear "statesmanlike" or to put the Soviets in a bad light[121]—there was definitely a limit on how far such initiatives could have been pursued. Even though the Cold War pressures of space had abated somewhat, the basic definition of space policy as the focus of U.S.–Soviet rivalry was still in place, and there was little chance that the president could move too far away from it.

Indeed, Kennedy did speak frequently of space competition, sometimes even mixing such references in with his statements on cooperation. The campaign position paper referred to above, for example, opened with a lengthy comparison of the two countries' programs. In an article written during the 1960 campaign, Kennedy adopted a tone strikingly similar to that heard earlier during the debate on the Space Act:

> Control of Space will be decided in the next decade. If the Soviets control space they can control earth, as in past centuries the nation that controlled the seas dominated the continents. . . . [W]e cannot run second in this vital race. To insure peace and freedom, we must be first.[122]

Such rhetoric continued after he took office. In fact, almost all of his public statements on the subject—with the notable exception of his appeals for cooperation—were laced with images of "racing," "winning," and competition.[123]

Inasmuch as Kennedy's statements on the subject are contradictory (as well as at odds with internal administration documents), it cannot be known for sure how he really felt about the space program prior to April 1961. In fact, at least one observer felt that, of all issues, he probably "knew and understood least about space."[124] Irrespective of whatever personal feelings the incoming president had on the matter, it is clear that many other central officials held to the view of space policy as Cold War competition. Thus, as had happened with the Eisenhower White

House, this particular conceptualization of space exploration came to dominate most Kennedy administration discussions of policy.

To take only the most prominent example, in marked contrast to Glennan, James Webb, the new NASA administrator, frequently raised the specter of Soviet space supremacy, particularly in discussions with the president.[125] At their first meeting (on March 21, 1961), Webb's presentation opened with a reference to "closing the gap caused by Russian successes," and concludes that "[w]e cannot regain the prestige we have lost without improving our present inferior booster capability, and doing it before the Russians make a major break through [sic]."[126]

For his part, Kennedy appeared fairly cautious. Although he did restore much of NASA's FY 1962 funding, the new figure was still 40 percent below the agency's request. Moreover, it is clear that the president was unsure how to proceed regarding the man-in-space program. Some have speculated that he wished to take a more in-depth look at manned flight beyond Project Mercury, and was waiting for Vice President Johnson to take over as chair of the National Aeronautics and Space Council to study the matter.[127]

Thus, as of April 1961, the Kennedy administration appeared to have a variety of philosophies with regard to space policy. How he would have chosen among them, however, will never be known. As happened in 1957, external events were once again about to intervene.

Conclusion

Few cases demonstrate the utility of the concepts presented in chapter 2 better than the development of U.S. space policy in the late 1950s. The analysis presented here has shown that, despite the steady growth, evolution, and (one hopes) maturation during its early years, the fundamental nature of NASA—indeed of the entire space program—was firmly established and set in place in just nine months. Between the announcement of *Sputnik I* in October 1957 and Eisenhower's signing of the National Aeronautics and Space Act the following July, the U.S. federal government's role in exploring space had been totally redefined. Despite the efforts of President Eisenhower and his senior officials to prevent it, and later of President Kennedy (perhaps) to change it, by the time NASA went into operation in October 1958, a preponderance of the nation's political leaders, and a large portion of the general public, regarded the space program predominantly in terms not just of national prestige, but of national security.

All of its potential uses notwithstanding, no other way of viewing space could ever have brought about such vast and sweeping changes in government organization and policymaking. None of space technology's other benefits—not communications or weather forecasting or as an aid to navigation, important though these applications may be—would ever have been sufficient to justify placing a Space Council in a potentially commanding position in the executive branch.[128] It is impossible to imagine how space exploration's scientific, economic, or social contributions by themselves would have led a congressman in 1958 to declare, without a trace of irony, that the director of NASA "is destined to be one of the most influential men in this country."[129] These and other such events were due exclusively to what was perceived to be the space program's—and NASA's—role in defending the country, specifically that of "meeting the Soviet challenge."

As chapter 2 also noted, how a policy is defined (or redefined) largely determines the manner in which it is carried out. With regard to space, the most obvious impact—and the one that was to have the most lasting effect—was in the organization of NASA itself. Under Eisenhower's original "science and surveillance" definition, the programs could be (and, in the case of military reconnaissance, were required to be) kept small and closed. However, once space policy was seen as an integral part of national survival—in effect, a sort of surrogate for the military—housing it in a large, centralized bureaucratic organization was inevitable. Eisenhower may have told Glennan that he wanted a "sensible" space program (and, for this president, "sensible" meant small), and may have sought to slow its growth by cutting its budget just before leaving office,[130] but as long as the majority in Congress believed that NASA held the key to "national survival," its continued growth was assured. In addition, the high priority assigned to its mission, along with its heightened visibility, provided NASA with a secure and reliable base of political support. For an agency charged with developing and operating new and untried (not to mention expensive) technologies, such backing was essential, particularly during its earliest years.[131]

Understanding how problem definition affects policymakers' approaches to public issues also provides some additional ammunition to an ongoing debate in space history. Some writers—particularly those who regard Project Apollo as an expensive mistake—have speculated about what might have happened if the USSR had not "beaten" the United States into orbit in 1957. What if, for example, von Braun had been allowed to launch his satellite in 1956, or *Vanguard* had come before *Sputnik*? Most believe that the American space program of the 1960s and beyond would have been much more "sound" and "logical" had it not been forced into a Cold War–fueled "race."[132]

It does seem apparent that the sort of political opportunities described in this chapter would not have been present had there been no "*Sputnik*s crisis," and thus no "rush" to "catch" the Soviets. There would almost certainly have been no Apollo acceleration (see next chapter) and perhaps no NASA. Simply knowing what would not have happened, however, says little about what would have. Given that, as seen in the last chapter, Eisenhower defined space primarily (if not exclusively) as a military intelligence issue, and that even the scientific satellite project had been approved largely because of its value to the intelligence program, it is far from clear that the administration would have gone forward with any new space initiatives, particularly after *Vanguard*'s cost overruns.

On the other hand, Eisenhower's later statements to the effect that the United States. faced "not a temporary emergency . . . but a long-term responsibility"[133] comes across today, particularly in light of NASA's experiences in the years after Apollo (see chapter 6), as rather perceptive. Moreover, as perhaps the only man in Washington at the time who did not accept the new definition of space policy, Eisenhower did serve to moderate—at least at the policymaking level—some of the panic resulting from the *Sputnik*s. For better or worse, four years later, when the next "space panic" would find another occupant in the White House, the results would be very different.

In addition, it is clear that the early history of NASA and the space program was very much driven by the political and personal goals of each of the principal participants. Scientists in virtually all fields saw *Sputnik* as an opportunity to obtain more funding for their research. Members of Congress (particularly Democrats) saw it as a chance to enhance their political status and power. Lyndon Johnson saw it as a potential stepping-stone to the presidency. Once again, it should be stressed that this general observation is not intended to be cynical. Rather, the point is that U.S. space policy during this time was being shaped by a number of powerful forces and inducements.

Finally, the creation of NASA reflects practically every type of reorganization strategy laid out in chapter 2. It was, first and foremost, a means for carrying out new policy initiatives. In bringing together the disparate (and at times competing) space programs described in the preceding chapter, it served the goals of policy integration and coordination. As congressional debate makes clear, the establishment of NASA (and, to a lesser degree, the National Aeronautics and Space Council) also played an important symbolic role, "demonstrating" to the Soviet Union, the West European allies, nonaligned nations, and the public at large American "determination" and "resolve." Each of these elements would continue, in one way or another, to play a role in the agency's development for the next several years.

In short, NASA was, on a variety of levels, a political creation, designed and animated by an unusually large number of participants, each with their own goals and objectives. As the next chapter will show, for the next few years, most of the political forces acting upon the agency would be mutually reinforcing, that is, in general agreement concerning its scope, size, and particularly its mission. Unfortunately for NASA, this near unanimity would not last for long.

5
Mission Advanced

Broad declarations of government policy, such as Lyndon Johnson's declaration of "war" on poverty or George H. W. Bush's vow that the Iraqi invasion of Kuwait would "not stand" (or, for that matter, his son's declaring war on terrorism) are, in and of themselves, really nothing more than broad statements of intention. They have real meaning only to the extent that they are accompanied by some sort of specific actions, such as—in the case of the War on Poverty—the creation of concrete programs and projects, or—as in Desert Storm (or Enduring Freedom)—direct orders to military commanders.[1] Of course, even when there is widespread agreement among policymakers and the public at large about the general goals, there may still be considerable controversy over the particular initiatives selected to achieve them.[2]

For most U.S. citizens (and many elected officials) in 1961, the tangible proof that their country had become "first" in space—that is, that the promises made repeatedly throughout the late 1950s and early 1960s by Presidents Eisenhower and Kennedy, as well as the 85th, 86th, and 87th Congresses had finally been fulfilled—would be for the United States to "beat" the USSR to the next "milestone," putting a man into space. Thus, despite warnings by experts like Jerome Wiesner, Americans by early 1961 had come to view NASA's Project Mercury as the primary (if not the only) government program specifically designed to meet the stated goal of taking the "lead" in space away from the Soviets. Once again, however, the United States was destined to come in "second"; and once again, its second place finish was to have far-reaching consequences.

Losing Round Two

Shortly after 1:00 AM (Washington time) on Wednesday, April 12, 1961, U.S. radar detected a rocket launch from the Soviet space center in Central Asia. One hour later, Radio Moscow announced that Yuri Alekseyevich Gagarin, a 27-year-old major in the Soviet air force, had just become the first man to fly in space. His spacecraft, *Vostok I*, carried him on a single orbit of the earth at a maximum altitude of more than 200 miles in a little less than 90 minutes. After bringing the *Vostok* out of orbit, he ejected from the spacecraft at around 23,000 feet and landed safely by parachute in a pasture near the Volga River.[3]

Many, if not most, authors tend to treat the Gagarin flight as a sort of *Sputnik* redux (Walter McDougall's prize-winning book even refers to it as a "second *Sputnik*"[4]), at least as far as American and world reaction was concerned.[5] Viewed in a larger context, however, it can be argued that the overall impact of *Vostok* was actually far greater than that of *Sputnik*. Indeed, its effects—as filtered through the political and institutional conditions of the times—are still felt strongly today.

To be sure, there were a number of similarities between the two events. As they had in 1957, the Soviets immediately sought to maximize the flight's propaganda value. Government leaders and other commentators touted *Vostok* as further proof of their technological prowess and the virtues of socialism. Nikita Khrushchev called it "the greatest triumph of the immortal Lenin's ideas."[6] The Central Committee of the Communist Party declared that the flight "embodied the genius of the Soviet people and the powerful force of socialism.[7] *International Affairs*, a Moscow journal, stated that Soviet space accomplishments were the result of

> the specific features of Socialist society, in its social structure, its planned economy, the abolition of exploitation of man by man, the absence of racial discrimination, in free labor and the released creative energies of peoples. Our achievements in the field of technology in general and in rocketry in particular are only a result of the Socialist nature of Soviet society.[8]

For *Pravda*, the official Communist Party newspaper, the flight represented nothing less than "the global superiority of the Soviet Union in all aspects of science and technology."[9]

Once again, a number of Americans seemed to take this claim very seriously. A *Washington Post* editorial published the day after the Gagarin flight echoed the *Pravda* declaration almost exactly, stating that "many persons will of course take this event as new evidence of Soviet superiority."[10] A NASA scientist remarked, "Wait until the Russians send up three men, then six, then a laboratory, then start hooking them together and then send back a few pictures of New York for us to see."[11] Congressman Victor L. Anfuso, a member of the House Science and Technology Committee, worried, "Lord knows where the Russians will be [in ten years] . . . and whether America will still be in existence."[12]

Vostok also rekindled the fears that Soviet triumphs in space would translate into a strategic advantage in the Third World and elsewhere. A *New York Times* editorial published five days after the flight worried that "[t]he neutral nations may come to believe that the wave of the future is Russian,"[13] and journalist Hanson W. Baldwin wrote, "Even though the United States is still the strongest military power and leads in many aspects of the space race, the world—impressed by the spectacular Soviet firsts—may believe we lag militarily and technologically."[14] Along these same lines, a U.S. senator noted:

> In the world stadium, nations are carefully watching the contest between the major protagonists of freedom and communism in space exploration—the United States and the Soviet Union. Although we didn't plan it that way, this is, indeed, a real space race. According to experts, the cumulative scientific-technical value of our space accomplishments far exceeds that of the Soviet Union. Nevertheless, we cannot ignore nor underestimate the psychological impact which Russian firsts in space have had upon the minds of the world. . . .
>
> [T]he task now is to first, predetermine the next major accomplishment in space; and second, as soon as possible, set up the timetable so that the United States can get the maximum benefit not only from the accomplishments themselves, but also from the great psychological impact of firsts in the space race.[15]

And Congressman Overton Brooks declared:

> I know that every member of the House Committee on Science and Astronautics is convinced that a great deal more depends upon the success of our national space effort than a simple race for scientific achievement between the United States and Russia. Right now the esteem in which the United States is held in the eyes of the world is dependent upon what we do in space as it is dependent upon few other areas of national endeavor. This is so whether we like it or not. The security of our Nation, in fact our very survival, our economic and material well-being, depend in no small degree on what we are able to accomplish in this space program. This is true of both the civilian and military aspects of the program.[16]

The potential impact of Gagarin's flight in the Third World had clearly not been lost on Soviet leaders, who, even while the mission was in progress, took full advantage of the fact that a human being makes a much better spokesman than a beeping satellite. Gagarin later said, for example, that during the flight he had been "thinking of our Leninist Party, our Soviet Motherland,"[17] and that during reentry he was singing "the Motherland hears, the Motherland knows."[18] His single orbit of the earth had also allowed him to transmit "revolutionary greetings" to countries in South America and Northwest Africa.[19] A cartoon captioned "In Tune with the Times—Africa!" from a 1961 Soviet publication depicts a smiling cosmonaut in orbit (aboard a craft labeled *Vostok*, but which bears little resemblance to the actual vehicle), waving to an African citizen, who has evidently just broken his shackles and is reaching heavenward.[20] If the press reaction to the flight in some developing countries is any indication, many were impressed.[21]

Finally, the American "response" once again came across as somewhat feeble.[22] Three weeks after *Vostok*, NASA launched Alan B. Shepard aboard *Freedom 7* on the first mission of the Mercury series. Unlike Gagarin, who had actually gone into orbit, Shepard's flight followed a parabolic, "suborbital" trajectory that lasted only 15 minutes. *Vostok* weighed almost five times as much as the Mercury capsule. Gagarin experienced 90 minutes of weightlessness; Shepard just 5.[23]

Such an evident "gap" led, once again, to numerous calls for the United States to "catch up." Representative James Fulton of the House Science and Technology Committee complained at a hearing the day of the flight, "I am tired of coming in second best all the time,"[24] and President Kennedy reportedly said a few days later, "there's nothing more important" than catching up.[25]

Number Two, Trying Harder

Despite all of these common features,[26] the fact is that Yuri Gagarin was flying over a very different political world than that traversed by *Sputnik* three and a half years earlier. The depth of these changes, and their impact on the subsequent conduct of American space policy, can best be illustrated by drawing on some of the concepts introduced in chapter 2.

Problem Definition

As galling as *Sputnik* might have been, Americans in 1957 could at least console themselves with the idea that they never knew that they were in a "race" to begin with.[27] According to this view (which was much encouraged by the Eisenhower administration[28]), the Soviets had, in effect, stolen a march on the United States. Put another way, *Sputnik* had been launched in a political vacuum. The idea that the space program's purpose should be to meet and overcome a Soviet challenge was not automatic; indeed, a number of officials, particularly those in the Eisenhower administration, actively resisted it.

By 1961, however, this definition of space policy was firmly in place. Despite the advice of numerous officials (including some from the incoming Kennedy administration) against placing too much emphasis on the manned portion of the space program, the enormous public attention focused on Project Mercury, and particularly on the seven Mercury astronauts, inevitably gave rise to the impression that America was aiming to be the first to put a man in space. In other words, to most Americans, "losing" this latest "race" was very different than not being the first to orbit a satellite. This time, it appeared to many that the United States had lost a direct, head-to-head competition.[29] In some respects, therefore, *Vostok* represented an even worse defeat for the United States than did *Sputnik*. And this time, the Americans could not accuse the USSR of sneaking up to the starting line.

In other words, the primary political issue raised by *Sputnik* was the direction of the U.S. space program. All of the speeches, exhortations, and criticism directed at Eisenhower and his officials were all geared toward getting the administration to accept a particular vision of space policy (and all that went with it). After *Vostok*, the question had changed to one of sufficiency: Kennedy now seemed to be open to the charge that he had not "done enough" to stave off this latest "defeat."

Reorganization

Along with settling upon the definition of the program, Congress and the administration expended a great deal of time and energy during the 1957–1958 period in setting up a number of space-related organizations and institutions, in both the executive and the legislative branches. By 1961, those agencies, committees, and councils had been up and running for more than two years. Thus, all of the effort that had gone into the general mobilization that followed *Sputnik* now went directly into the development of specific programs and proposals (see below). What is especially important to note here is that each of these organizations had a strong interest in expanding the program.

For example, the House of Representatives' Committee on Science and Technology had already scheduled a round of hearings when the *Vostok* launch was announced. The committee heard from Edward Welsh of the National Aeronautics and Space Council on April 12, and from James Webb, Hugh Dryden, and Robert Seamans of NASA on April 13 and 14.[30] None of these three organizations—the House Committee, the National Aeronautics and Space Council, or NASA—even existed three years earlier.

Moreover, these new organizations and agencies provided government officials and the public at large—and particularly the media—with a focal point for their attention. Following his appearance before the House Committee on April 14, Seamans described stepping out of the hearing room into a "blinding light of television cameras" and a sea of reporters demanding to know how the United States intended to respond to the Gagarin flight.[31] The officials clearly knew that they were being watched. The Science and Technology Committee hearing on April 13, for example, was moved to the much larger Cannon caucus room in order to accommodate the committee's "friends" in "the press, radio, and television."[32]

Politicians and Goals

By far the most significant differences between *Sputnik* and *Vostok*, however, was with respect to the major political actors. This was, of course, most obvious at the top. As the last chapter suggested, President Kennedy's approach to space policy in the first few months of his term bore a close resemblance to his predecessor's. This may have been, as some have suggested, due to his relative lack of interest (at that time). Even so, taking into account each man's overall temperament, their philosophical outlooks, and relative political position, it is difficult to think of two more different leaders. Eisenhower was older, more cautious about government spending, and more committed to stability. He was serving his second and final term at the time that *Sputnik* was launched. Kennedy was a younger man who wished to convey an impression of vigor and vitality. He had campaigned on a pledge to "get America moving again." He was, by all accounts, less resistant to large government programs than was Eisenhower. Finally, he was still in his first term—indeed, his presidency was barely three months old at the time of the Gagarin flight.

Not surprisingly, Eisenhower's reaction to Kennedy's Apollo decision (see below) was decidedly negative. Writing to a friend, the former president characterized his successor's proposal as "almost hysterical" and a bit immature.[33] On numerous occasions, in public and private, Eisenhower sharply criticized the moon program as a waste of the nation's resources.

Some congressional Republicans, however, seemed to part company with their former president's views (at least for a while). Having been on the receiving end of constant Democratic criticism in the aftermath of *Sputnik* and throughout the 1960 presidential campaign, many in the GOP saw *Vostok* as a prime opportunity for political payback. For example, confronting Webb at the April 13 meeting, Congressman Fulton roundly criticized Kennedy's decision in March to agree to only part of NASA's budget request. Webb had no choice but to defend the president's actions, even though he privately agreed that the budget was inadequate.[34]

Another event that may have been an influence in the Apollo decision occurred just one week after Gagarin's flight. An armed contingent of Cuban exiles, trained and supported by the CIA, failed in their attempt to overthrow Fidel Castro's regime. The defeat of the U.S.-backed guerrillas at the Bay of Pigs represented a serious foreign policy and domestic setback for the still-new Kennedy administration. Despite the efforts of the president and others to blame the affair on Eisenhower's policies, it was beginning to appear that, in author Tom Wolfe's words, "the 'New Frontier' was looking like a retreat on all fronts."[35]

There is considerable disagreement as to how much any of this influenced Kennedy. W. Henry Lambright's recent biography of James Webb, for example, notes that there is no direct evidence linking the Bay of Pigs and Apollo.[36] Lyndon Johnson wrote in 1971 that "he [Kennedy] never gave the least indication that in any of our discussions that he thought there was any relationship."[37] On the other hand, Arthur M. Schlesinger does draw a direct connection between the invasion, Gagarin, and the moon challenge.[38] Curiously, virtually all of the attention in this regard is focused on the possible effects on Kennedy personally. There has been relatively little investigation how the Bay of Pigs may have affected the rest of White House staff, members of Congress,[39] or the public at large. As will be seen below, many individuals had input into the Apollo decision, and any of them could have been motivated by the defeat in Cuba, or at least may have become more receptive to a proposal that promised to reassert U.S. leadership.

Finally, the 1960 election brought into the White House a number of individuals who were far more willing to define space policy in Cold War terms than most Eisenhower officials had been. First and foremost among these, of course, was Vice President Lyndon Johnson. As the last chapter noted, Kennedy had already requested that the National Aeronautics and Space Act be amended to allow Johnson to serve as chairman of the National Aeronautics and Space Council. Although the council had been largely inactive during the Eisenhower administration, the new president clearly intended to put the vice president's expertise to use.[40] Johnson, who absolutely hated the vice presidency,[41] could at least settle for some position of responsibility over a highly visible issue (and one that had given him some measure of national prominence). He would later play a key role in persuading reluctant members of Congress to support NASA's expansion.[42]

Other officials seemed equally disposed to view space as a "race" with the Soviets.[43] Secretary of State Dean Rusk told the Senate Space Committee that he feared a "misunderstanding" among other nations "concerning the direction in which power is moving and where long-term advantage lies."[44] Defense Secretary Robert S. McNamara, in a memo to Johnson concerning the acceleration of the space program (see below), noted:

> What the Soviets do and what they are likely to do are . . . matters of great importance from the viewpoint of national prestige. Our attainments constitute a major element in the international competition between the Soviet system and our own.[45]

As chapter 2 pointed out, however, policymakers can use a single policy position to pursue multiple goals, and there is reason to believe that McNamara was no exception. He had already been overseeing cutbacks in defense spending that were upsetting many in the aerospace industry.[46] A larger space effort would easily fill the gap left by these cuts.

In short, for the first time in the young history of the Space Age, the leadership in the Congress and most of the officials in the executive branch were, for the most part, on the same side of the issue. Given this fact, as well as his own political situation, it was all but inevitable that Kennedy would begin to look for the next major spaceflight "milestone." Or, more precisely, to try to select a milestone sufficiently distant that the United States would have the time it needed to "catch up" at long last.

Going to the Moon[47]

As of early 1961, NASA's proposed timetable for human spaceflight called for a circumlunar mission by 1970, with a landing at some unspecified time in the future.[48] The Gagarin flight, however, quickly made that timetable—which had represented a sort of compromise between the cautious Eisenhower administration and the more ambitious space advocates—obsolete.[49] On April 20, 1961, Kennedy sent a memo to Vice President Johnson asking for a survey on where the United States stood "overall" in space. Item number one, however, makes it very clear where the president's priorities lay:

> Do we have a chance of *beating the Soviets* by putting a laboratory in space, or by a trip around the moon, or by a rocket to land on the

moon, or by a rocket to go to the moon and back with a man? Is there any other space program which promises dramatic results *in which we could win?*[50]

In announcing Johnson's assignment to the press the following day, Kennedy added that he was seeking to identify a project in which, "regardless of cost," the United States could be a "pioneer."[51]

In preparing his report for the president, Johnson received input from numerous individuals in and out of government. Far from challenging the basic premise underlying Kennedy's request, most participants embraced it openly, even to the point of continuing the "racing" imagery. Wernher von Braun, for example, spoke of the United States having a "sporting chance" of landing a probe on the moon or sending a three-man crew around the moon ahead of the Russians.[52] He also claimed that the United States had an "excellent chance of beating the Soviets to the first landing of a crew on the moon" since this would require a rocket that neither side yet possessed.[53] The McNamara memo referred to above spoke of the need to "match the Soviets in all areas of international competition."[54]

Johnson's initial reply to Kennedy, in a brief memo dated April 28, continued this theme:

> This country should be realistic and recognize that other nations, regardless of their appreciation of our idealistic values, will tend to align themselves with the country they believe will be the world leader—*the winner* in the long run. Dramatic accomplishments in space are being increasingly identified as a major indicator of world leadership.
>
> If we do not make a strong effort now, the time will soon be reached when the margin of control over space and over men's minds through space accomplishments will have swung so far on the Russian side that we will not be able to catch up, let alone assume leadership.[55]

It goes on to identify a lunar landing as an "achievement with great propaganda value" and as a goal in which "we may be able to be first."[56]

The final report was transmitted to the vice president on May 8 (the Monday after Alan Shephard's flight). It suggests "four principal reasons" for undertaking projects in space: scientific research, commercial enterprises, defense, and national prestige. In an odd sort of backhanded compliment, the report conceded that the United States "is not behind" in the first three categories: "Scientifically and militarily we are ahead. We consider our potential in the commercial/civilian area to be superior." The USSR, on the other hand, leads "in space spectaculars which bestow great prestige." In laying out the logic behind this observation, the report makes an admission that is rather startling in its frankness:

> All large scale space projects require the mobilization of resources on a national scale. They require the development and successful appli-

cation of the most advanced technologies. They call for skillful management, centralized control and unflagging pursuit of long range goals. Dramatic achievements in space, therefore, symbolize the technological and organizing capacity of a nation.

It is for reasons such as these that major achievements in space contribute to national prestige. Major successes, such as orbiting a man as the Soviets have just done, lend national prestige *even though the scientific, commercial or military value of the undertaking may by ordinary standards be marginal or economically unjustified.*[57]

Just in case any doubt as to the primary purpose of U.S. space policy remained, this section of the report concludes:

The non-military, non-commercial, non-scientific, but "civilian" projects such as lunar and planetary exploration are . . . part of the battle along the fluid front of the cold war [sic]. *Such undertakings may affect our military strength indirectly if at all*, but they have an increasing effect upon our national posture.[58]

In other words, like congressional debate over the original Space Act, this report presents—and immediately discounts—the other justifications for developing spaceflight technology. In fact, it goes even further, suggesting that highly visible "space spectaculars" can be justified even if they prove to be wasteful from an economic, scientific, or military point of view. Overall, the report provides the most direct statement yet of space policy definition during this period.

The culmination of all of this activity was Kennedy's famous "Urgent National Needs" speech (described as an unprecedented second State of the Union Message[59]), delivered before a joint session of Congress on May 25, 1961. The most well-known part of the address, in which the president declares a lunar landing before 1970 as the primary goal of the United States in space, strongly suggests that the country was largely out to recapture the prestige it had lost to the USSR:

I believe that this nation should commit itself to achieving the goal, before this decade is out, of landing a man on the moon and returning him safely to the earth. No single space project in this period will be more impressive to mankind as it makes its judgment of whether the world is free or more important for the long-range exploration of space; and none will be so difficult or expensive to accomplish.[60]

It is in the other, less often quoted parts of the speech, however, that Kennedy makes it very clear who is the primary target of the policy. Indeed, he referred to the Soviets and their space program twice, clearly casting them as America's principal adversaries in both space and politics, and openly declaring that they were the primary reason for the United States going to the moon:

> [I]f we are to win the battle that is now going on around the world between freedom and tyranny, the dramatic achievements in space which occurred in recent weeks should have made clear to us all, as did the Sputnik in 1957, the impact of this adventure on the minds of men everywhere, who are attempting to make a determination of which road they should take.
>
> Recognizing the head start obtained by the Soviets with their large rocket engines, which gives them many months of lead-time, and recognizing the likelihood that they will exploit this lead for some time to come in still more impressive successes, we nevertheless are required to make new efforts on our own. For while we cannot guarantee that we will someday be first, we can guarantee that any failure to make this effort will make us last.[61]

Given that many of its members had been calling for just such a response to the *Vostok* flight, it is hardly surprising that Congress overwhelmingly approved Kennedy's proposal, voting an immediate 50 percent increase in NASA's budget.

Of course, as had happened a few years earlier during the debate over the Space Act (see last chapter), government officials did from time to time mention the other benefits of space technology. As already noted, the May 8 report made brief references to scientific and commercial applications. One seldom noted feature of the Urgent National Needs address is that, in addition to the lunar mission, Kennedy asked Congress to approve an additional $125 million for continued development of communication and weather satellites.[62] He would also, on other occasions, note that his accelerated space program promised more than just national prestige:

> In the national interest, the United States must build the capacity to advance the most modern science and technology to the utmost, and extract from it the wealth of benefits it holds for this country's freedoms, economy, professions, and standard of living.[63]

Nevertheless, as was the case in 1958, it is highly unlikely that even these "wealth of benefits" would have been enough to justify the billions of dollars Project Apollo would ultimately cost,[64] or that the potential economic or scientific results by themselves would have persuaded Congress to support the program. Indeed, a survey of House members taken in early 1962 by the magazine *Aviation and Space Technology* found that most thought the expense of the space program "questionable except when viewed from the standpoint of the Cold War."[65]

In short, most policymakers—including by 1961 key members of the executive branch—were now committed to a policy definition that conceptualized spaceflight (particularly that involving humans) as a critical element in U.S. national security policy. Kennedy himself put the matter rather bluntly in 1963 when he said that the U.S. goal in going to the moon was "not only our excitement or interest in being on the moon, but the capacity to dominate space, which . . . I believe is essential to the United States as a leading free world power."[66]

Like any political debate over budget increases for (conventional) defense spending, defining space policy in a way that makes it essential to "national survival" allowed Apollo's supporters to answer—or, in some cases, ignore—criticisms of the program's high costs, which began in earnest in late 1961 and continued for the rest of the decade. In his public statements, for example, Kennedy would sometimes raise the threat of another "new, dramatic [Russian] breakthrough"[67] "which could affect our national security"[68] when asked about the costs of the moon program. Indeed, for the rest of his life (with one major exception to be discussed below) the president described his space policy almost exclusively in Cold War terms. In his press conferences, reports, and public speeches, he practically never mentioned the U.S. program without also referring to Soviet capabilities. His rhetoric was constantly laced with images and metaphors of racing: he would, for example, concede that the United States was running "second" or "behind" the USSR,[69] but at the same time expressed confidence that it would soon "catch up" and "be ahead . . . where . . . we ought to be."[70] On the day before his assassination, in a speech in San Antonio, Texas, the president proclaimed:

> I think the United States should be a leader. A country as rich and powerful as this . . . should be second to none. . . . [W]hile I do not regard our mastery of space as anywhere near complete, while I recognize that there are areas where we are still behind . . . this year I hope the United States will be ahead.[71]

Vice President Johnson, as usual, was somewhat more direct. He once told a group of congressmen who were expressing some reluctance on Apollo, "[W]ould you rather have us be a second-rate nation or should we spend a little money?"[72]

A similar pattern can be seen in official response (or, rather, the lack of it) to the growing disapproval among professional scientists. Most members of the science and engineering community tended—not unreasonably—to view space policy simply as one among many of the nation's R&D programs. They believed—again, not unreasonably—that the size, pace, and funding of such programs ought to be based on their potential scientific or technological return. Although they were not yet organized in any meaningful way (that would come later), a number of prominent scientists and engineers were highly critical of Apollo, since the scientific returns could not possibly justify the costs: anything astronauts could do on the moon could be done more cheaply (and more safely) by automated probes.[73] One of the more notable critics was Vannevar Bush, the principal architect of postwar U.S. science policy. Although he had initially seen *Sputnik* as a "wake-up call" for Americans who had grown too smug,[74] Bush favored a more balanced approach to space R&D, somewhat along the lines of the Eisenhower administration. Throughout the early 1960s, he wrote numerous articles decrying Project Apollo, and even sent a letter to this effect directly to Kennedy himself (through Webb, whom he had known from the Truman administration).[75] He was by no means alone. A "straw poll" conducted by the editor of *Science* magazine in 1963 of 116 scientists "not connected by self-interest" to NASA found only 3 in favor of the manned space program.[76]

Chapter 2 noted, however, that policymakers pay attention to outside information only to the extent that it conforms to their way of conceptualizing the issue. Since most public officials had long ago stopped seeing the U.S. space program as "science"—the May 8 report to the vice president stated explicitly that space missions conferring national prestige did not need scientific (or, for that matter, economic or military) justification—scientists' objections were simply dismissed as irrelevant (there is no indication that Kennedy even replied to the Bush letter[77]).

Finally, chapter 2 described how policymakers' definition of public problems guides them in their selection of proposed solutions. In this case, the "solution" meant not only deciding on a goal—landing on the moon by 1970—but also on the means of getting there. The overriding objective beating the Soviets (which, since no one knew precisely the USSR's plans or emerging capabilities, was taken to mean simply that the appropriate technology had to be developed as rapidly as possible) was to play a major role in shaping the technologies that made up the Apollo Transportation System.[78]

Mission planners originally conceived of three different approaches to getting astronauts on to and off of the lunar surface: direct ascent (a nonstop flight with a single, very large rocket), earth orbit rendezvous (EOR; launching the lunar landing craft from near-earth orbit), and lunar orbit rendezvous (LOR; the landing craft would leave from and return to lunar orbit). Direct ascent was quickly discarded, as there seemed to be no chance of building and testing a booster of required size anytime in the near future.[79] Many officials, including Kennedy's science advisor, favored EOR because it lent itself more directly to other space-related activities beyond the lunar missions.[80] It would therefore better serve the general, long-run interest of the program. Unfortunately, it also presented a number of technical problems—most notably, large-scale construction and fueling a spacecraft in orbit—that engineers were not certain could be addressed successfully within the president's timetable.

LOR, the option eventually selected, did present some additional risk, in that some of the more difficult parts of the mission, such as rendezvous and docking, would be taking place in lunar orbit, far from any hope of rescue. Nevertheless, even supporters of EOR conceded that the approach had the virtue of being the one most likely to meet Kennedy's target date,[81] which it did. On the other hand, it also represented a form of "lock-in" for NASA. As chapter 2 noted, an agency may find its future actions constrained by the scale, cost, or complexity of technical systems or technological infrastructure selected at an earlier point in time. While the elements of EOR might (after all, this is all speculative) have readily lent themselves to missions beyond Apollo, the hardware used in LOR proved to be highly specialized. As a result, NASA was forced after the lunar missions to redesign its space transportation system, at considerable expense.[82]

The Seeds of Evolution?

Chapter 2 stated that issue evolution—a shift in the patterns of government support for a policy or program—can be influenced by a number of factors, such as interest group activity, external events, and general political change. A number of observers (including some writing at the time) have suggested that, in effect, the Cold War–based definition of U.S. space policy—and thus its commitment to the Apollo program—began changing as early as 1963.[83] While fixing a precise date when support for a given policy begins to change is a rather difficult task, it is clear that many of the elements that chapter 2 described as contributing to issue evolution were indeed present in the 1962–1963 period.

To begin with, President Kennedy was coming under increasing pressure to revise his position on the lunar program. Although (as seen earlier) some Republican members of Congress had, in the first few days after the Gagarin flight, attacked the president for not investing enough in the space program, by 1963 GOP representatives and senators had taken to criticizing him for spending too much. In May, for example, the Senate Republican Policy Committee released a report urging that Apollo be scaled back and that the money be redirected to— using a phrase that would be heard more and more frequently in the years ahead— "problems here on earth."[84] That fall, Congressman Louis Wyman of New Hampshire sought a $700 million reduction ($550 million of which was to come from the moon program) in NASA's FY 1964 budget. The proposal was soundly defeated (132–47), but the fact that it was presented at all is an indication that opposition was beginning to develop.

More serious, from Kennedy's point of view, was the growing criticism coming from prominent members of his own party. Senator J. William Fulbright of Arkansas emerged during the latter half of 1962 as a staunch opponent of Apollo. His primary objection, he said, was not the lunar goal itself, but rather the end-of-the decade timetable, which added considerably to the cost of the program.[85] He was joined by fellow liberal Senators Joseph S. Clark (Pennsylvania) and Ernest Gruening (Alaska), both of whom (like Fulbright) wanted to increase funding for social programs. Fulbright's attempt to cut NASA's 1964 budget by 10 percent fared somewhat better than Wyman's similar effort in the House, losing by only 10 votes (46–36).[86]

In addition to growing (albeit still relatively small) domestic discontent, there were a number of international events that served if not to reduce Cold War tensions at least to reassure Americans about their status in the world. In October 1962, the United States successfully pressured the Soviet Union to withdraw nuclear missiles that it had placed in Cuba. To whatever extent the Bay of Pigs debacle the previous year had played any role in Kennedy's (or, for that matter, Congress' or the public's) thinking about space, the American "victory" in the Cuban Missile Crisis may have provided an effective antidote. In addition, the signing of a limited Nuclear Test Ban Treaty with the Soviets several months later appeared to set a (somewhat) more cooperative tone in superpower relations.

Finally, although its validity has been the subject of much debate, the president was presented with evidence that there might not even be a moon race after all. In a 1962 report (in response to a proposal to accelerate Apollo; see further discussion below), Kennedy's budget director noted that "there is no evidence . . . that the Russians are actually developing a large booster . . . or the rendezvous techniques" required to mount a lunar mission.[87] This same view was aired publicly the following July, when the British astrophysicist Sir Bernard Lovell wrote a letter to NASA describing his recent visit to a number of Soviet space installations. During the course of his discussions, he learned that, due to the technical difficulties involved, the USSR had no plans to land a cosmonaut on the moon anytime in the near future, but would be open to a joint mission with the United States. As might be expected, the letter (the Bureau of the Budget [BOB] report had not been made public) created an immediate controversy.[88] Some officials, clearly still operating under the prevailing Cold War definition of space policy, warned that the admission to Lovell could have been a "Soviet trick" to distract the country from its vital mission.[89] For his part, Kennedy seemed to brush off the new information:

> The kind of cooperative effort which would be required for the Soviet Union and the United States to go to the moon together would require a breaking down of a good many barriers of suspicion and distrust and hostility which exist between the Communist world and ourselves. There is no evidence as yet that those barriers will come down, though quite obviously we would like to see them come down. . . . I would welcome it, but I don't see it as yet, unfortunately.[90]

Nevertheless, there are some small indications that the prevailing definition of space policy was beginning to change, and that even the president himself (bearing in mind that there may never be any way to prove this definitively) may indeed have been reevaluating the Apollo commitment. It is useful to recall at the outset of any such discussion that Kennedy had not embraced the Cold War vision of the program in his first months in office. As seen in chapter 4, much of his early (i.e., pre-Gagarin) rhetoric was aimed at promoting space cooperation, and he had even ordered his science advisor to begin drafting proposals for joint U.S.–Soviet space projects. He was also prepared to propose a number of cooperative science and technology programs, including space research, directly to Chairman Khrushchev during their 1962 summit in Vienna. Although the offer was not made formally (due to the generally contentious nature of the meeting), the president did raise the possibility of a joint lunar mission during a lunch conversation. Khrushchev did not make any commitment at that time, but he did accept a proposal, made by Kennedy a few months later, to join the United States in a limited number of space projects.[91]

Certainly the most dramatic event in this vein was the president's speech before the United Nations in October 1963, where he publicly called for a joint mission to the moon, using language that actually seemed to contradict the justification he himself had given for Apollo two years earlier:

> [I]n a field where the United States and the Soviet Union have a special capacity—in the field of space—there is room for new cooperation, for further joint efforts in the regulation and exploration of space. I include among these possibilities for a joint expedition to the moon. Space offers no problems of sovereignty. . . . Why, therefore, should

man's first flight to the moon be a matter of national competition? Why should the United States and the Soviet Union, in preparing for such expeditions, become involved in the immense duplications of research, construction, and expenditure? Surely we should explore whether the scientists and astronauts of our two countries—indeed of all the world—cannot work together in the conquest of space, sending some day in this decade to the moon not the representative of a single nation, but the representatives of all of our countries.[92]

In view of all of this activity, particularly the UN speech, some scholars have concluded that Kennedy's "true" preference was for space cooperation with the Soviets—or at least to avoid extending the Cold War into that realm—but that he felt the need to adopt a more competitive posture in the face of *Vostok* (and perhaps the Bay of Pigs).[93] One—somewhat less charitable—author has gone so far as to claim that the moon challenge was the result of the president being "rattled by momentary crises."[94] Either assessment, if correct (and, once again, it should be stressed that there is no direct evidence either way), would suggest that a bit of distance from the events of 1961, in combination with some policy successes (and perhaps growing confidence in his reelection prospects), might have led Kennedy to consider looking for a face-saving way out of the moon race.[95]

Another clue as to the president's thinking was his decision (whether it would have been technically feasible is another matter) not to speed up the Apollo program. In late 1962, R. Brainard Holmes, NASA's Apollo project director, had called for moving the projected date for the first lunar landing to late 1966.[96] This proposal, and particularly the large budget increases it would have required, prompted Kennedy in November to ask the Bureau of the Budget for an overall review of the space program (referred to earlier), and to pursue the matter directly with Webb.

The BOB report, dated November 13, and a follow-up letter from Webb, written two weeks later, were generally supportive of the Apollo program as it then existed, but recommended against any further acceleration. The BOB memorandum began by noting that Apollo's projected costs were rising faster than had been anticipated and, as discussed above, that there was no direct evidence that the USSR was engaged in a moon program of its own. It then moved, however, into a discussion of U.S. space policy that, in both language and reasoning, departed significantly from (if not directly contracted) the arguments employed by McNamara, Johnson, and others in formulating the lunar program just 18 months before:

> The special attention given to the manned lunar landing program has sometimes obscured the other program objectives being pursued by NASA. Perhaps the most important are the programs for scientific investigations in space, in which the United States from the start has been recognized as the world leader, which have intrinsic value, which have been the focus of significant programs of international cooperation, and which, in some cases . . . can provide spectacular achieve-

ments with some of the same popular appeal as manned space flights. Less costly, but most important, are the programs directed at developing practical applications of space technology, chiefly in the meteorological and communications fields.[97]

The report further notes:

> NASA takes the view that the importance of maintaining the proposed general level of effort in the "other" areas is so great that if any reduction were to be made in [the] $6.2 billion budget request, *it should be applied at least in part to the manned lunar landing program*, in order to maintain a "balanced" total program.[98]

Webb's letter to Kennedy generally echoes this theme. He notes that "NASA has many flight missions, each directed toward an important aspect of our national objective," adding:

> Although the manned lunar landing requires major scientific and technological effort, it does not encompass all space science and technology, nor does it provide funds to support direct applications in meteorological and communications systems. Also, university research and many of our international projects are not phased with the manned lunar program, although they are extremely important to our future competence and posture in the world community.

He concludes that "the manned lunar landing program, although of highest national priority, will not by itself create the preeminent position we seek."[99]

Thus, for the first time since *Vostok* and the establishment of the lunar landing goal, members of the administration—including the director of NASA—are taking the position (at least in internal discussions) that other areas of space R&D, if not quite as high a priority as Apollo, are nonetheless essential components of the agency's overall mission. Such a view is in stark contrast to the sentiments expressed in documents like the vice president's May 8, 1961, report. It is even more striking to compare Kennedy's original April 20, 1961, memo to Johnson, which asked, "Are we working 24 hours a day on existing programs[?] If not, why not?" and "Are we making maximum effort?"[100] with the BOB report, which concedes that rejecting Holmes's proposal "will be criticized in some quarters as representing slightly less than a maximum effort."[101] Even so, Kennedy evidently found the new logic persuasive. Apollo was kept on its original schedule, and Holmes left NASA soon after.

Of course, it does not necessarily follow that issue evolution (if that is in fact what was occurring here) happens overnight, or that it will take place among all relevant policymakers simultaneously. It should be noted, for example, that although both documents described above extol the virtues of non-Apollo space projects to an unprecedented (at least since mid-1961) degree, they are still careful to place those projects in the same general Cold War context as before (i.e., that they will enhance national prestige and the U.S. "world posture," have popular appeal, will help the United States achieve "preeminence" in space, and so forth).

Moreover, it is quite clear that at least some administration officials did not share this "evolving" view. In April 1963, Kennedy asked Johnson, in his capacity as chair of the Space Council, to provide him with still another report to provide "a clearer understanding of a number of factual and policy issues relating to the National Space Program."[102] This time, it was the president himself who raises the issue of spaceflight's "other benefits":

> What specifically are the principal benefits to the national economy we can expect to accrue from the present, greatly augmented program in the following areas: scientific knowledge; industrial productivity; education, at the various levels beginning with high school; and military technology?

It has been suggested that Kennedy's purpose in making this request was not to begin a major reevaluation of the program, but rather to seek additional justification (and arguments) for it. A significant point in this regard is that the memo was addressed to the National Aeronautics and Space Council (as opposed to, say, BOB), which was certain to give a positive response.[103] Moreover, even if Kennedy had been feeling pressured by criticism over NASA's budget, he nevertheless made it clear where his priorities lay, regardless of his stated interest in the program's "additional benefits": the memo specifically asks Johnson to identify places where the program could be "reduced . . . in areas not directly affecting the Apollo program."[104]

In his response, the vice president dutifully noted, "It cannot be questioned that billions of dollars directed into research and development . . . will have a significant effect on our national economy." The report cites the usual advances in space science, communications, and weather forecasting, as well as a number of specific improvements in industrial technology and materials sciences, an "augmentation" in the supply of trained technical manpower, and "greater strength for the educational system." In his conclusion, however, Johnson not only retains the logic of the May 8, 1961, report, but essentially returns to the rhetoric of his Senate days, making it abundantly clear what he believed truly justifies U.S. space policy:

> There is one further point to be borne in mind. The space program is not solely a question of prestige, of advancing scientific knowledge, of economic benefit or of military development, although all of these factors are involved. Basically, a much more fundamental issue is at stake—whether a dimension [sic] that can well dominate history for the next few centuries will be devoted to the social system of freedom or controlled by the social system of communism. . . .
>
> We cannot close our eyes as to what would happen if we permitted totalitarian systems to dominate the environment of the

earth itself. For this reason, our space program has an overriding urgency that cannot be calculated solely in terms of industrial, scientific, or military development. The future of our society is at stake.[105]

This view seemed equally entrenched among many members of Congress, as can be seen in the response to Kennedy's UN speech proposing a joint lunar mission. As it happened, the president's address coincided with congressional debate on NASA's FY 1964 budget, in which the Apollo program was coming under increasingly heavy criticism. Opponents of the project immediately made use of what they saw as the "illogic" of the administration's position, asking, for example, "[i]f there is going to be military value in a trip to the moon, how is that going to be possible if it is done jointly with Russia?"[106]

Apollo's supporters in Congress, clearly still operating under the Cold War definition, felt that they had been betrayed. Albert Thomas and Olin Teague each sent Kennedy strongly worded letters of protest, the latter expressing "disappointment" over the proposal, and asking if "this national goal [of being the first nation to place a man on the moon] has changed."[107] On the floor of the House, Teague argued forcefully for continued Apollo funding, but also declared that "for goodness' sake I hope we never reach the point of trying to get into a capsule and go to the moon together."[108]

Attempting to understand whether Kennedy's position on the moon program really had changed was not made any easier by his response to Thomas's letter (there is no indication that he ever replied directly to Teague), dated September 23. The president stated that seeking out cooperative ventures with the Soviets had been administration policy from the beginning (which, as has been seen, was true), as had efforts to build up U.S. capabilities in space. Thus, he concluded, the "great national effort" of Apollo and the "stated readiness to cooperate" were not at all in conflict, but were rather "mutually supporting elements of a single policy."[109]

What the president really intended for Apollo (as well as what he actually meant in his letter to Thomas) will never be known with any certainty. He was assassinated in Dallas just a month and a half after the controversy erupted, and up to the end, his actions on the question appeared contradictory. On the one hand, he ordered NASA to begin preparing "specific technical proposals" for a joint lunar expedition.[110] At the same time, he never mentioned the idea in public again, except to note in answer to a question at a press conference that he had not received a response from the Soviets,[111] and for the most part returned to calling for U.S. space leadership (as in his last speech, delivered in San Antonio).[112]

There is, however, another way of interpreting Kennedy's proposal that is still consistent with the Cold War definition of space policy.[113] As has already been shown, superpower competition in space was more than a simple head-to-head contest of technological prowess (although that was obviously a major component). The larger issue behind the space race of the late 1950s and early 1960s was which political system could successfully present itself to the rest of the world as the more "progressive," "forward-looking" society. Kennedy *may* therefore have reasoned that, by making such a sweeping, unprecedented offer to move into the newly dawning Space Age together, the United States could claim that it, although still at that point "behind" in terms of raw technology, was truly the more "visionary" country.[114] Paradoxically (as was so much of foreign and domestic policy during the Cold War), "cooperation" can be seen as another, more subtle form of competition.

Whatever the case, it is highly unlikely that a joint lunar mission could have taken place during the 1960s, even if Kennedy had lived. First, there is some question about its technical feasibility. Robert Gilruth, director of NASA's Manned Spacecraft Center (later called the Johnson Space Center) reportedly told a National Rocket Society meeting in September 1963, in response to a question about the Lovell letter, that he "trembled at the thought of the integration problems" in merging American and Russian space systems."[115]

Second, the Russians never accepted the offer. In fact, they never made any official response at all beyond Khrushchev's remark on November 1: "What could be better than to send a Russian and an American together, or, better still, a Russian man and an American woman."[116] Given the level of secrecy surrounding the USSR program in the early 1960s, it is difficult to imagine—even if they had agreed to the proposal—that the Soviet government would have, in the end, been willing to share enough information for true cooperation to occur.

Finally, perhaps the most serious obstacle (and the one most relevant to the present discussion) is that it would have been against the law. On October 10, the House passed (125–110) an amendment to the NASA appropriation preventing the agency from participating in a lunar landing project jointly with "any Communist, Communist-controlled, or Communist-dominated country."[117] The language was later softened by the Senate to prohibit cooperation on a lunar program with any "other country" without the consent of Congress,[118] a restriction that would be included in the NASA appropriation for the next three years. As it happened, however, the congressional action was to have little impact on the conduct of the program. The new president, Lyndon Johnson, was somewhat obligated to support (at least verbally) his predecessor's call for cooperation, and in January 1964, he did duly receive the NASA report that Kennedy had ordered.[119] Not surprisingly, he did little to advance the UN proposal. His view of space policy—for the first few months of his presidency, at any rate (see next chapter)—continued to be that the United States should use the program to demonstrate its space leadership:

> Our plan to place a man on the moon in this decade remains unchanged. It is an ambitious and important goal. In addition to providing great scientific benefits, it will demonstrate that our capability in space is second to no other nation's.[120]

Thus, as of the beginning of 1964, the Cold War definition of NASA's mission was still firmly in place.

Conclusion: The High Point?

From this point until the *Apollo 11* landing in July 1969, the agency would encounter few, if any, problems (of a political nature, at any rate) in completing

Kennedy's mandate. Indeed, to the casual observer, NASA's and the space program's prospects during 1964 and 1965 could hardly have seemed brighter. First, it was growing rapidly. The number of full-time workers grew from 10,000 in 1960 to more than 34,000 in 1965.[121] The number of employees working on a contractual basis rose from 36,500 in 1960 to more than 370,000 (i.e., by more than a factor of 10) in 1965.[122]

Second, its political backing appeared stronger than ever (the few dissenters notwithstanding). Congress continued to approve large budget increases: between 1962 ($1.8 billion) and 1963 ($3.67 billion), NASA appropriations more than doubled, and rose to $5.1 billion in 1964 and $5.25 billion in 1965. Lyndon Johnson, one of NASA's principal architects and its most ardent champion, had not only succeeded to the presidency, but, in the 1964 election, was given one of the largest electoral mandates in modern American history. Even the general public—despite the criticism from some congressmen and members of the science and engineering community—generally stood behind the program. Polls taken during the 1964 election found that substantial majorities favored moving ahead with the lunar landing.[123]

Third, while some Americans might have complained about the high costs of the program, no one could deny that at least the country's space capabilities were obviously improving. In 1959, NASA's first full year of existence, the U.S. success rate in launching spacecraft was only 50 percent (nine successes, nine failures).[124] By 1962, that figure had improved to 82 percent and by 1965 had risen to 93 percent.[125] Moreover, it began to look as though the United States was actually moving ahead of the USSR, at least as far as sheer quantity was concerned. Despite starting out "behind" at the end of the 1950s, by 1965 the United States had had a total of 270 successful launches compared to 130 for the Soviet Union. It even seemed as though the United States was finally "catching up" in manned spaceflight, the one area that the Russians had dominated since 1961. After completing the Mercury series in 1963, NASA launched its first 2-man spacecraft, *Gemini 3*, in March 1965. The agency set a number of milestones during the three Gemini flights that year, including the first manual orbital maneuvering by a spacecraft (*Gemini 3*), the first extravehicular activity (i.e., "space walk") by an American (Ed White from *Gemini 4*), and the longest human spaceflight up to that time (*Gemini 5*, almost eight days).[126] Probably the most important achievement of this period was the launch of the first Saturn-I rocket on January 29, 1964. For the first time, the United States had a booster that could lift heavier payloads than the USSR could.

Despite these feats, as well as the claims by some that the Russians were not in fact going to the moon,[127] the space race—at least in terms of the image presented to the public—showed no signs of slowing down. In October 1964, six months before *Gemini 3*, the USSR launched the world's first multiple crew, three cosmonauts, aboard *Voskhod 1*. The headline in *Pravda* the following day read "Sorry, Apollo!" and an accompanying article, using language almost identical to that from the days of *Sputnik 1* and Yuri Gagarin, ridiculed the idea that the United States would ever catch up:

Now such prophecies from the Americans can bring forth an ironic smile. The gap is not closing, but increasing. This is natural . . . the so-called system of free enterprise is turning out to be powerless in competition with socialism in such a complex and modern area as space research.[128]

The official American response to *Voskhod* clearly echoed the concerns that dated back to *Sputnik* and Gagarin:

> The flight of the multi-passenger ship *Voskhod* demonstrated a continuing Soviet ability to mount flights on a scale surpassing that of the U.S. The capacity of the Soviets to launch larger spacecraft in manned flight remained the negative factor affecting foreign opinion of U.S. space activities.[129]

The following March (five days before the first Gemini flight and two months before Ed White's "space walk"), a Russian cosmonaut, Alexei Leonov, performed the world's first extravehicular activity from *Voskhod 2*. In short, although the U.S. program clearly was making progress, Americans who had always feared Soviet exploits in space still seemed to have reason to be afraid (and to continue to support NASA's programs).

Nevertheless, there were still signs that major political change was coming. Although the agency's budget was growing steadily, Congress was beginning to give NASA less than it asked for. The appropriation for FY 1964 ($5.1 billion), for example, was $500 million below the agency's request. The FY 1965 budget ($5.25 billion) was trimmed by less than $100 million, but this was to be the agency's high-water mark: funding for NASA (as measured in constant dollars) would never be this high again. More important, policymakers—including those who might have been considered the most likely to rally to the agency's side—chose not to get involved in one of the most important events, from the agency's point of view, to take place during this period: the selection of the missions that were to come after completion of Project Apollo.

At first glance, it might seem strange that a public agency for which matters were going so well would encounter such difficulties. As the next chapter will show, however, the reason NASA began, in the words of one historian, its "eclipse"[130] at this time is actually rather simple: the Cold War definition as a way of conceptualizing space policy was beginning to lose its political persuasiveness. This, in turn, raised the question of what definition could be found to replace it.

PART III

SECOND MISSION?

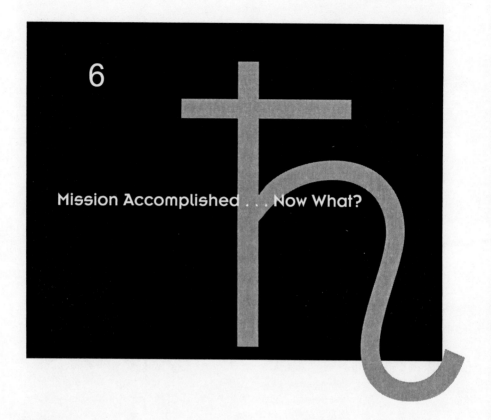

6

Mission Accomplished... Now What?

As the Apollo 11 command module *Columbia* splashed down in the Pacific Ocean on July 24, 1969, two statements appeared across the 20 × 10 foot video screen in NASA's Mission Control center in Houston. The first was the famous challenge issued by President John F. Kennedy eight years earlier:

> I believe that this nation should commit itself to achieving the goal, before this decade is out, of landing a man on the moon and returning him safely to the earth.

Below it was the declaration:

> Mission Accomplished, July 1969[1]

Most observers date the scaling back of the U.S. space program from this moment. "And then," says one author, "it was all over,"[2] or, as another put it, "the television sets around the country began to flick off."[3] Somewhat more soberly, a 1981 report for the House Space Science and Applications Subcommittee notes simply that once this "primary goal" had been achieved, "the pace of the U.S. program slowed considerably."[4] What these authors—not to mention the celebrants in Mission Control—failed to realize, however, is that the true goal of the space program, NASA's *real* mission, had actually been achieved some years earlier.

The Bigger They Are . . .

Between 1960 (the final Eisenhower budget) and 1965, NASA's funding grew by *900* percent. When adjusted for inflation, its 1965 budget represented a *thirteenfold* increase over that of five years earlier. After hitting this peak, however, the agency went through 10 straight years of spending cuts (as measured in constant dollars). In fact, it was not until 1982, the year after the first space shuttle flight, that the NASA budget for space activities (again in constant dollars) matched that of 1962, the year of John Glenn's first orbital flight.[5] Of course, as chapter 2 noted, public issues come and go, political priorities change, and the budget will always be shifting as a result. Still, it is difficult to imagine another public issue that matches the speed and scale of NASA's roller-coaster ride of the 1960s;[6] significantly, the closest approximation would be to the pattern of military spending just before and after a major war.

To be sure, the reductions began slowly. NASA's appropriation for FY 1966 (approximately $5.18 billion) was only about $75 million less than the previous year, a decline of around 1.5 percent. Still, it was the first time since the beginning of Apollo that the agency had received less than the year before, and it clearly made James Webb nervous. His efforts to secure an increase for FY 1967 (to $5.3 billion) were rejected by the Bureau of the Budget,[7] with the result that NASA funding fell below $5 billion (to $4.97 billion) for the first time since 1964.

By now seriously alarmed, Webb informed President Johnson that 1968 was a "critical" year, and that anything less than a $6 billion NASA budget would mean the "liquidation" of many of the capabilities the agency had built up (such as production lines for the giant Saturn booster; see below), the loss of numerous future mission opportunities, and widespread demoralization throughout the agency.[8] Despite these warnings, the president signed a FY 1968 appropriations bill that included a $500 million *cut* for NASA. This was followed several months later by another $500 million reduction for FY 1969, dropping the agency's budget below $4 billion.

Under these conditions, NASA was scrambling simply to keep its current programs on schedule. It is therefore hardly surprising that few officials were giving much attention to—much less making serious plans toward—what types of missions might follow the completion of Project Apollo. In early 1964, Johnson asked NASA for "a statement of possible objectives beyond those already approved."[9] Webb, however, was wary of the agency being put in the position of proposing its own missions. As he saw it, the more proper approach (and one less politically risky for the agency) was for the nation's political representatives, that is, its elected officials, to tell NASA what they wanted it to do, much as Kennedy had done with Apollo.[10]

Accordingly, in early 1965 (several months late), Webb sent the president a Future Programs Task Group report outlining a number of projects—including the use of robot probes to explore Mars, continued exploration of the moon, and use of the Apollo-Saturn technology for earth applications—that the agency could undertake after Apollo, but made no specific recommendations.[11] In his cover letter, Webb reminded Johnson, "More than in most areas, major decisions on space require a broad consensus."[12] The president, however, refused to take the hint, and made no comment on the recommendations.

The agency took a somewhat similar approach with Congress, listing the types of missions that would become "feasible" after Apollo, but making no clear statement of priorities. A 1965 presentation on "Advanced Manned Missions" made before the House Subcommittee on Manned Spaceflight, for example, noted NASA could, in the future, undertake missions of 2–3 months' duration in space science and "earth-oriented applications" in near-earth orbit, as well as extended lunar exploration missions of up to 2 weeks.[13] As was the case with Johnson, however, Congress never made a formal commitment to any of these proposals.

Throughout the mid-1960s, Webb tried time and again to get some sort of major post-Apollo commitment out of the president, to no avail. Matters became particularly acute as development of major elements of the Apollo Transportation System neared completion. Some NASA facilities, most notably the launch complex at Cape Canaveral in Florida and Mission Control at the Manned Spacecraft Center in Houston, would be fully engaged with Project Apollo up until the very end. Others, however, such as the Marshall Space Center in Huntsville, which oversaw the construction of the giant Saturn V booster, would have completed their assignments long before. Similarly, the private contractors who manufactured Apollo's components had orders for only a specified number of items. These companies would need to be told well in advance whether they could expect future orders. As private businesses, the only way they could afford to keep open the lines dedicated to the production of Apollo hardware would be if they knew that the government was committed to making further purchases.

Working against this impending deadline, Webb hoped to move the president to a decision on a post-Apollo program by warning him in the summer of 1966 that the agency was faced with no other choice but "to accelerate the rate at which we are carrying on the liquidation of some of the capabilities we have built up."[14] Indeed, one of the reasons Webb was so anxious to secure approval for the Apollo Applications Project, an orbiting space laboratory built largely from existing hardware,[15] was to keep the Apollo production lines open as long as possible.[16] This was, however, simply staving off the inevitable. By August 1968, with no new major program forthcoming, Webb had no choice but to order that Saturn rocket production be shut down.[17] It was to be one of his last acts as NASA administrator: the following month, he announced that he would retire from government service on October 2, his 62nd birthday.[18]

The NASA administrator was not the only one frustrated by Johnson's silence. Beginning in mid-1965, Congressman Olin Teague, chairman of the Manned Spaceflight Subcommittee, had been pressing NASA on its post-Apollo goals. During budget hearings in March 1966, Teague reacted angrily to the news—delivered by an equally frustrated Webb—that the president intended "to hold open for another year the major decisions on future programs." "To me," said the congressman, "it is like telling a child that we are going to make you crawl another year before you can walk."[19] The following July, the committee directed NASA "to report to the Congress, no later than December 1, 1966, its recommendations on possible major national space objectives," including estimated costs, benefits, and its composition in terms of manned and unmanned elements. When the deadline came, however, all the committee received was a 2-page letter with Webb's name typed at the bottom, but signed by Robert Seamans, citing uncertainty as to "what the president will approve" as a reason for not providing the detailed analysis requested.[20]

In short, by 1968,

> After a heady decade of uninterrupted hiring, building, and dreaming great dreams of far-reaching exploration, the American space program is gearing down to a slower pace and a less certain future. . . . The growing feeling in the space establishment is that once astronauts have landed on the moon, they will have no other place of significance to go for several years because of sharp budget cuts. These cuts have trimmed to the bone all preparations for future missions. It is as if the astronauts are heading for a dead end on the moon.[21]

Live by the Sword . . .

As might be expected, at the root of the agency's decline was a substantial—and, in some cases, rather sudden—change in NASA's political support. Nowhere is this more obvious than with regard to President Johnson. Indeed, given how much time and effort he expended on behalf of NASA and the space program while in the Senate and as vice president, it is rather startling to realize how little this issue seemed to hold his attention once he became president. As biographer Robert Dallek has noted, Johnson's own account of his presidency, a 600-page volume, contains only 17 pages about space, almost all of which (14 pages, to be exact) deals with events prior to 1963.[22] During an interview with Walter Cronkite in 1969, he admitted "quite frankly" to doing much more for the program in "'57 and '58 and '59 and '60, and up to 63, than I did after I became President."[23]

Moreover, the president (the same politician who, just a few years earlier had asked some wavering congressmen, "Would you rather have us be a second-rate nation or should we spend a little money?"[24]) was now acceding to—and in some instances even initiating—substantial cuts in the NASA budget. As noted above, Johnson's BOB blocked Webb's effort to secure a funding increase for FY 1967, imposing a $200 million reduction instead. Some members of Congress, however, wanted to slash space spending even further. Wisconsin Senator William Proxmire had proposed cutting the agency by an additional $1 billion. In a clear indication of how the political tides surrounding the program had shifted, Webb felt compelled to seek the assistance of Illinois senator Everett Dirkson, the Republican leader, to head off Proxmire's proposal (which was defeated). "It is never an easy thing," he later told the president, "to decide the time has come to ask for help from the minority leader."[25]

The $500 million reduction in NASA's FY 1968 budget came about in part through a political deal. Earlier in 1967, the administration had projected that the budget deficit for FY 1968 would be around $8 billion. By that summer, however, it became clear that the true figure would be much higher—close to $30 billion—and Johnson felt as though he had no choice but to ask Congress for a 10 percent tax increase. Congress did comply with the president's request, but the House of Representatives also decided that the deficit reduction should be accomplished by both tax increases and cuts in federal spending, and reduced NASA's appropriation that year accordingly. Despite Webb's protests that this would prove "the straw that break's the camel's back,"[26] Johnson felt he had no choice but to go along. In approving the measure, he noted:

> Under other circumstances, I would have opposed such a cut, [but] the times demand responsibility from us all.
>
> I recognize—as also must the Congress—that the reduction in funds recommended by the House Appropriations Committee will require the deferral and reduction of some desirable space projects. Yet, in the face of present circumstances, I join with the Congress and accept this reduction.[27]

Privately, the president told Webb that he did not "choose to take one dime from my space appropriations for this year," but had to agree or risk losing the tax bill.[28] Although Johnson told the NASA administrator that he hoped "to make up for this" the following year,[29] in 1968 the administration on its own proposed reducing the agency's budget by $250 million, and ultimately accepted the $500 million cut imposed by Congress.

To be sure, Johnson did continue to talk about, and act on behalf of, the program. As some authors have noted, however, by late 1964 Johnson's rhetoric around space exploration had mellowed considerably.[30] Although he did continue, from time to time, to talk about the program in comparative—if not necessarily competitive—terms (see below), for the most part his speeches and press conferences were no longer laced with vivid and apocalyptic images (recounted throughout the previous two chapters) of communist domination of the heavens.

Indeed, the president often spoke of the need for more cooperation in space.[31] In 1965, for example, following a telephone conversation with returning Gemini astronauts, Johnson said, "We do hope and we do pray that the time will come when all men of all nations will join together to explore space together, and walk side by side toward peace."[32] Similarly, while speaking with employees of the Manned Spacecraft Center in Houston, the president invited those living under communism "to open your curtains, come through the doorways and the walls that you have built, and join with us to walk together toward peace for all people."[33]

Indeed, Johnson at times seemed to embrace a view of space that transcended traditional superpower politics and explicitly rejected national competition. "We have," he declared in 1965, "no need for arms races or moon races."[34] The following year, when accepting the Goddard Trophy, the president claimed that "[t]he true significance of space is the story of victory over the forces of nature."[35] And once, during a press conference, he asked:

> As [man] draws nearer to the stars, why should he also not draw nearer to his neighbor? As we push even more deeply into the universe, we must constantly learn to cooperate across the frontiers that really divide the earth's surface.[36]

This sea change in Johnson's sentiments (if not his style) was far more than just rhetoric. He actively sought to act on these newly found cooperative impulses. In 1964, he sent NASA's Hugh Dryden to Geneva to seek out new opportunities for space cooperation. The trip ultimately came to nothing, due to an apparent reluctance on the part of the Soviet Union to enter into any new agreements. Nevertheless, the president persisted. In late 1963, while the body of President Kennedy was still lying in state in the White House, the president consulted with UN ambassador Adlai Stevenson about U.S. policy on weapons in space.[37] This conversation led the president, in 1966, to open the negotiations at the United Nations that ultimately produced the Outer Space Treaty of 1967, which banned space-based nuclear weapons, required the assistance and safe return of astronauts who landed in another country, and prohibited claims of sovereignty over the moon or other celestial bodies.[38]

In addition to Johnson himself, there were other influential White House officials who were beginning to take a more skeptical view of NASA and the "space race." The State Department, for example, urged the "defusing" of the space race and the "stretching out" of "costly programs aimed at the moon and beyond" so as to free up resources for other needs, such as "foreign aid, domestic needs, [and] scientific efforts in other areas."[39] Donald F. Hornig, Johnson's science advisor, took issue with some of Webb's budget requests and his growing sense of alarm, especially later in the decade.[40]

It was Budget director Charles L. Schultze, however, who proved to be the most constant critic of Webb's assessments of space policy and (particularly) the latter's fears that continuing funding cuts would mean "losing" the moon race.[41] In fact, in 1967, Schultze even went so far as to urge the president to abandon the 1969 lunar landing goal. Noting that "we may fail in any event" to make the end-of-the-decade deadline,

> [w]hy not make a virtue out of necessity? It would be better to abandon the goal now in the name of competing national priorities, than to give it up unwillingly a year from now because of technical problems.[42]

Two points seem clear here. First, unlike the near unanimity over Apollo that had reigned just a few years earlier, the White House now seemed more sharply divided over the future (and even the current) course of the program. Second, the fact that Schultze would make such a direct challenge to NASA's requests, and particularly the lunar landing goal, strongly suggests that Johnson himself may have been feeling ambivalent about the program. While budget directors in particular are prone to oppose most costly programs, it is highly unlikely that Schultze would be quite so forthright in his assessment of NASA and Webb if he thought that both had the president's unequivocal support.

Congress seemed to be cooling on the program as well. Although NASA still had strong support among some of the more important members of both the House and the Senate, it is clear that the institution as a whole having second thoughts about an aggressive U.S. posture in space. An internal White House survey of congressional attitudes taken in late 1966 revealed that most members favored keeping Apollo on track, but wanted to reduce NASA's budget by cutting back on most post-Apollo programs (which was, in fact, precisely the course Johnson followed).[43] In addition, the president appears to have read Congress' mood correctly in 1967 when he accepted its budget cuts in exchange for his proposed tax surcharge.

Finally, public support seemed to be evaporating. Virtually all public opinion polls taken during the mid- to late 1960s showed significant differences—sometimes on the order of two to one—between the percentage of respondents who favored reducing government spending on space and those who supported increasing it.[44] In short, it appears as though, except for NASA and a relatively small group of space enthusiasts, most people—public officials as well as ordinary citizens—wanted to see a smaller space program.

Conventional Explanations

Of course, it is not especially helpful to explain the space program's decline by simply saying that it had lost its earlier political support. Obviously this begs the fundamental question of why public, congressional, and administration support had eroded so precipitously. There are, however, no shortage of ideas on this topic.

"Antiscience" Attitudes

One common observation, for example, is that the space program became caught up in the general distrust toward scientists and engineers that had become part of the American social fabric by the mid-1960s. According to this view, the environmental movement (prompted by such books as Rachel Carson's *The Silent Spring* and Barry Commoner's *Science and Survival*), the consumer protection movement (which also had a founding book, Ralph Nader's *Unsafe at any Speed*), protests over the Vietnam War (and DOD-funded research generally), and the growing realization that science and technology could not address fundamental problems like poverty and urban decay all came together to create a social and political environment that was far more hostile to the science and engineering community than that which existed just a few years earlier.[45] As one of the more visible (not to mention more expensive) R&D projects of the 1960s, the space program was especially vulnerable to this sort of criticism (e.g., "we can send a man to the moon, but we cannot . . .").

Vietnam and Other Issues

Another commonly held view is that government officials—beginning with President Johnson—were becoming "distracted," both politically and financially, by other issues, most notably the war in Vietnam.[46] It is known, for example, that the president was receiving constant reminders about the cost of the war from his budget director, who frequently cited "the continued fighting in Vietnam" as a reason for cutting NASA's budget requests.[47] Moreover, Webb clearly believed the war was a major factor in the president's behavior. In the August 1966 letter referred to earlier, he told Johnson that he "believed firmly in the actions you are taking . . . in Vietnam" and that he had serious doubts about involving the president in the budget disputes because he had "no desire to add to your burdens."[48] The Webb/Seamans letter sent to the House Science and Technology Committee in late 1966 also referred to the war as the source of NASA's "uncertainty" concerning the president's wishes.[49]

Administration officials were not the only ones who saw a relationship between the cost of the war and funding for the space program. In a speech on the floor of the House in 1966, for example, Congressman Teague complained that "the war in Vietnam has already forced a substantial reduction in the NASA budget for the coming year."[50] It should also be remembered that the $500 million cut for NASA that Johnson was forced to accept in 1967 had been ordered by Congress to help address a $29 billion deficit created, at least in part, by the war.

There is also reason to believe that, in addition to its cost, Vietnam was consuming large shares of the president's personal attention as well. Webb believed that toward the end of his presidency, Johnson had become "obsessed" with the war.[51] It is also known that during his last few years in office, the president began each day by reviewing casualty reports and other accounts of the ground action.[52]

Finally, it seems clear that substantial portions of the general public were beginning to see spaceflight as less important than issues like Vietnam, race, or the condition of American cities. A 1967 poll by the *New York Times* showed that the public ranked five other policy areas as more important than space.[53] As a 1968 *Newsweek* article concluded

> The U.S. space program is in decline. The Viet Nam war [sic] and the desperate condition of the nation's poor and its cities—which make spaceflight seem, in comparison, like an embarrassing national self-indulgence—have combined to drag down a program where the sky was once no longer the limit.[54]

"What Have You Done for Me Lately?"

Another—far more cynical—explanation relates back to the November 1957 memo from George Reedy telling Johnson that the *Sputnik* issue could "elect [him] president" (see chapter 4). If, as some have suggested, Johnson's interest in space after *Sputnik* was fueled largely by personal political ambitions, there would have been little reason to maintain the same level of support beyond the 1964 election (particularly in the face of growing opposition in Congress and among the general public).[55] With his landslide victory over Senator Barry Goldwater, Johnson had achieved everything he had ever wanted in public life. In short, it is possible that, from the president's point of view, NASA and the space program had simply outlived their political usefulness.

This might explain, for example, why he sought out Webb's advice on future programs in early 1964, only to show little interest in the information when it was presented to him early the following year (i.e., after the election). It would also account for the sudden shift in his way of speaking about space. Having reached the presidency, and no longer in need of the attention and visibility that his alarmist (and at times outright bellicose) rhetoric brought him, Johnson could now afford to sit back and adopt a more "statesmanlike" tone.

Disorganization at NASA

Finally, some authors have criticized NASA itself for the lack of any Apollo follow-on program. It has been argued, for example, that the agency did not have any sort of planning mechanism in place to respond to President Johnson's first (1964) request, and that subsequent efforts to put together such a panel became bogged down among all of NASA's various constituencies (manned spaceflight, applications, space science, etc.).[56]

According to this view, it is no coincidence that the general questioning about what the United States should do in space coincided with the greatest catastrophe in NASA history (until the *Challenger* accident nearly 20 years later). On January 27, 1967, three astronauts—"Gus" Grissom, Ed White, and Roger Chaffee—were killed when a fire broke out in their *Apollo I* spacecraft during a routine launch simulation. Subsequent investigations into the accident—along with Webb's apparent efforts (now generally regarded as ill-advised) to withhold from congressional investigators an internal report critical of North American Aviation, the prime contractor for the Apollo command module—led, for the first time, to NASA's competence being called into question.[57] Indeed, one editorial writer quipped that the initials in the agency's name really stood for "never a straight answer."[58]

◉ ◉ ◉

It is possible, of course, to take issue with each of these accounts of the space program's decline. To begin with, although there clearly was a sort of "backlash" against science and technology among sections of the general public (especially the so-called counterculture), it is far from clear that this had much of an effect on federal R&D policy (at least not until the early 1970s; see below). The rate at which overall funding for science and technology increased did slow somewhat toward the end of the decade (in fact, measured in constant dollars, it even decreased slightly in 1968 and 1969), but only one other field—to be discussed further below—experienced anything like NASA's reverses. Indeed, some agencies, such as the National Institutes of Health, received budget *increases* during this period.[59]

Vietnam certainly was expensive and, particularly as the decade progressed, politically divisive and time-consuming. Still, it is worth remembering that the war was started, at least initially, for the same general reasons as the space program: to meet the perceived threat of communism. On the face of it, there is little reason why one high-cost anticommunist program by itself ought to squeeze out another.

Although his enthusiasm clearly was diminished, it is simply not the case that Lyndon Johnson started ignoring the program once he became president. He frequently referred to space in his speeches, and was known to follow each flight of the Gemini program closely. It is said, for example, that Johnson personally pushed NASA into including a space walk on *Gemini 4*. The agency had originally intended for an astronaut simply to stick his head and shoulders out of the spacecraft. After Alexei Leonov's walk on *Voskhod 2*, however, Johnson reportedly told NASA, "if the guy can stick his head out, he can also take a walk. I want to see an American EVA [extravehicular activity]."[60]

Finally, given the lack of consensus in the administration and Congress, it hardly seems reasonable to have expected NASA to develop its post-Apollo mission on its own, particularly with regard to large-scale projects. Moreover, as has already been discussed, Webb was reluctant to act without clear signs of strong political support.

The most important problem with virtually all of these explanations, however, has to do with the idea of "decline" itself. Since most (although by no means all) space histories are written by space enthusiasts, it is not uncommon for such "explanations" to focus on the program's post-Apollo "decline," as though the unprecedented buildup that preceded it were perfectly normal. On the other hand, opponents of the Apollo program—for example, many members of the science and engineering community, Eisenhower administration officials, some contemporary critics, and so forth—would argue that NASA's long string of shrinking budgets (at least until the approval of the space shuttle program) require no explanation whatsoever: they represent a necessary period of "readjustment." For them, what "really" needs to be explained is the abnormally large spending on space that occurred earlier in the decade.[61] The discussion over the past few chapters has been based on the premise that both events—NASA's rapid expansion and its almost-as-rapid decline—do not follow the typical pattern of American politics. Thus, it is important that whatever explanation is advanced for the one should also be able to explain the other.

"De-Definition?"

In defining space as a Cold War issue, political leaders like Kennedy and Johnson (along with key members of Congress) charged NASA with developing an American space capability that matched—and would eventually surpass—that of the USSR. All of the projects undertaken during the early 1960s were supposed, in one way or another, to contribute to this basic mission.[62] To take only one example, the previous chapter quoted President Kennedy explaining how Apollo was justified "not only [by] our excitement or interest in being on the moon, but [by] the capacity to dominate space, which . . . I believe is essential to the United States as a leading free world power."[63]

As the last chapter also noted, that that goal had, for all practical purposes, been met by the mid-1960s, clearly, the United States was no longer "behind" the Russians. Following *Voskhod 2* in 1965 (see last chapter), the United States launched 10 consecutive manned flights without a single Soviet response.[64] When the USSR finally did place another man in space, *Soyuz 1* in 1967 (after almost a 2-year hiatus), the pilot was killed during reentry.[65] The first successful *Soyuz* flight did not take place until late 1968, just a few months before the first U.S. circumlunar flight.

A similar trend could be seen in the area of unmanned scientific probes. Although the United States certainly had its share of problems with projects like Ranger (unsuccessful until the seventh attempt in 1964), the USSR's difficulties were even worse. The Russian *Luna 9* (February 3, 1966) did beat the U.S. *Surveyor 1* (May 16, 1966) to the first soft landing on the moon, and its *Luna 10* became the first artificial object to orbit the moon on April 3, 1966, beating American *Lunar Orbiter 1* by four months,[66] but over this same period they also experienced six (announced) unsuccessful probes of Venus, five of Mars, and four of the moon.[67]

Last, but far from least, the successful tests of large boosters like the Saturn I-B (the precursor to the massive Saturn V that would carry the astronauts to the moon and place the *Skylab* laboratory in earth orbit) beginning in 1966 had finally given the United States what it had always lacked in its competition with the USSR: a rocket powerful enough to carry very large objects into space.[68]

Not surprisingly, public statements by government officials, as well as internal documents, show a steadily increasing level of confidence that the United States was roughly even with—if not ahead of—the USSR in terms of its space capability. The president, for example, declared in 1965:

> We were unmistakably behind. Some prophesied that America would remain behind, that our system had failed, that the brightness of our future had dimmed and would grow darker. But no such prophesies are heard today.[69]

Similarly, the following year he noted that "we haven't wiped out all of the deficiencies yet, but we have caught up and are pulling ahead."[70]

In private, the assessments were similarly optimistic. A 1966 memo from Ed Welsh to Vice President Hubert Humphrey observed that through the end of Project Gemini, the United States had logged nearly 2,000 man-hours in space (including the six Mercury flights) compared to 507 for the Soviets, as well as more that 11 hours of space walking compared to 20 minutes for the USSR.[71] Welsh also notes a number of "firsts" in the Gemini program (some also discussed in the previous chapter): the first manual orbiting maneuvering, the first controlled rendezvous in space, and the first docking of a piloted spacecraft to another vehicle. Neither of the last two had been accomplished by the Russian program by this point.[72]

A few months earlier, Schultze informed the president:

> There is only *one area* of space activity *in which the number of Soviet launches to date exceeds that of the U.S.* But, in this area, planetary exploration, we have had *far better success* with our limited number of attempts than the Soviets have had with their larger effort.
> In *every other area of launches we have a substantial lead.* . . .
> Evidence indicates that *they* [the Soviets] *are working on a launch vehicle which may . . . exceed . . . our Saturn V*, but there are indications that *they are at least a year behind* our current Saturn V flight schedules. This situation is a good deal different from that which existed in 1961 when the Russians were demonstrating a weight-lifting capability superior to anything we expected to have for several years and were clearly ahead of us in manned space capability.[73]

Accordingly, references by NASA or administration officials to the Soviet space program diminish significantly beginning in 1965. Whether reporting the agency's progress, celebrating its achievements, or discussing post-Apollo plans, the USSR no longer cast the long shadow over U.S. space planning that it had earlier in the decade.

For example, in a statement marking the 10th anniversary of NASA, President Johnson never mentioned the Russians or the space race at all. Instead, he notes that "we have seen space science and technology assume a high-ranking position in human affairs." He singles out for special mention the probes to the moon, Venus, and Mar, and the development of weather and communication satellites. The closest he comes to any sort of Cold War reference is toward the very end, where he notes that "[o]ur program has been conducted openly, in the sight of the entire world."[74] Moreover, as noted above, on a number of occasions the president explicitly rejected the idea of "competition" in space.

Finally, it began to appear that the one area most feared by American policymakers during the early days of spaceflight—world opinion, particularly in developing countries—was finally breaking in the direction of the United States. A 1965 U.S. Information Agency report notes that, whereas three-quarters of Nigerians surveyed in May felt that the USSR was "ahead" in space, by September more than half felt that the United States now had the "lead."[75] An administration official stationed in Bolivia during the *Gemini 4* flight in 1965 reported that every "transistor-radio-liberated peasant" was aware of Ed White's walk in space and now "worships our technology."[76]

Stated simply, given NASA's ongoing success, the apparent lack of progress on the part of the Soviets, and the general "de-mystification" of space technology on the part of the U.S. public, "the anti-Russian theme," in the words of astronaut John Glenn, "had worn out."[77] As a result, NASA was about to be retired from Cold War service.

To be sure, the Cold War itself was still very much in progress, and many Americans (as well as many public officials) still harbored deep-seated fears of the

Soviet Union. Even so, it appears that by the mid-1960s most policymakers (along with a significant portion of the general public) had stopped seeing accomplishments in space as being particularly relevant to the larger superpower conflict. For the most part, those who continued to express concern over the USSR in space (mostly conservative Republicans) spoke of a gap in military space applications.[78]

Perhaps the clearest evidence of this can be found in the experience of James Webb, who, in trying to stave off the ongoing budget cuts at NASA, played the "anti-Russian theme" almost continuously (as he had almost from the beginning; see chapter 4). From 1966 until he left NASA in 1968, Webb would always warn of some imminent Russian breakthrough in space, that they could still beat us to the moon, and so on. After the USSR successfully soft-landed *Luna 9* on the moon in 1966, for example, Webb tried to warn the president that the United States could still lose the lunar landing "race."[79] He also warned of the "political consequences" that would come from underestimating Soviet capabilities.[80]

Webb took a similar line with the Congress. In presenting NASA's proposed 1967 budget to the Senate Appropriations Subcommittee, he stated:

> The competition is still fierce, and we are not yet able to feel assurance that we will end up ahead in the option areas where the Russians are developing their strongest potential. A $5 billion budget level in the years ahead will not be adequate to develop and utilize the options we are now in the final stages of developing.[81]

In March 1968, he told the House Committee on Appropriations:

> During this period when we are reducing our efforts by one-third, the USSR is still increasing its effort. We must therefore face the probability that in the coming year, and those following, the Soviets will continue to demonstrate capabilities beyond those which we have.
>
> The hard fact we now face is that just as we have begun to catch up in large-scale booster operations . . . we are cutting back our program while the Soviets continue to advance.[82]

When advising Budget director Schultze on the language that Johnson should use when signing NASA's FY 1968 budget, which included $500 million in cuts (see above), but which did not even mention the USSR's program, Webb noted, "As to possible language about Russian activity . . . I regard this omission as one the President will regret. . . . The activity of the Russians is . . . spectacular and . . . calculated to show the world how far in front they are."[83]

In 1968, after the Soviet Union successfully sent their *Zond 5* spacecraft (carrying two turtles) around the moon and back for a safe (albeit rough) landing in the Indian Ocean, Webb called it

> the most important demonstration to date of all the capabilities required for operations around the earth and outward to the moon

and planets—in other words, all the capabilities for any purpose in space.⁸⁴

In a September 16 press conference announcing his retirement, he complained that the United States was still second to the Soviets in space, and was so because NASA's budget had been cut so sharply over the past three years.⁸⁵

Webb was by no means alone in these efforts. Ed Welsh informed the Senate Appropriations Committee in May 1968 that "the acceleration of the space program in the Soviet Union is much greater than ours . . . their technology progress [*sic*] is greatly increased."⁸⁶ Earlier in the year, Wernher von Braun told the House Science and Astronautics Committee:

> There are no signs that the Soviets are cutting back in their space program. Under these circumstances our national posture in this competition is already rapidly worsening as the strength of the Soviet thrust is affirmed again and again while our own space budget is shrinking year after year.⁸⁷

These and other such claims were generally met outside of NASA with emotions ranging from general unconcern to outright skepticism. As already noted, the Bureau of the Budget rejected Webb's fears out of hand. Members of Congress—neither in committee nor on the floor—responded to these assertions of Soviet ambitions in space in quite the same way as they had during the period immediately after *Sputnik* and *Vostok 1*. It also seems that even the president—who had once evoked fears of Americans "going to bed by the light of a Communist moon"—had reached the conclusion that the United States had finally developed an insurmountable lead and definitely would land men on the moon ahead of the Soviets.⁸⁸

There was still a possibility that the USSR might have beaten the United States to the moon. (The Soviets reportedly had planned a circumlunar mission that would have flown two weeks before *Apollo 8*, but canceled it due to problems with the booster rocket⁸⁹). Such an event would certainly have been disappointing, but it almost certainly would not have inspired the same level of anxiety and fear as had *Sputnik* and *Vostok*. Despite such serious setbacks as the *Apollo I* fire, the United States had, by 1966–1967, achieved the primary goal of its space program. If it could not say definitively that it was indeed "first" in space—due solely to the fact that so little was known for certain about the status of USSR space policy—it could with great confidence claim that it had "caught up." Although the moon landing itself would not occur for two more years, it is clear that the United States had met its "real" objective in space by 1967.

As suggested earlier, none of this meant that public officials in any way regarded the Cold War as being over. Indeed, superpower competition was still a driving force in a number of policy areas,⁹⁰ including some aspects of space. Even as NASA's budget was shrinking, DOD funding for space activities was actually going up. Between 1965 and 1969, the budget for space-based defense grew by nearly one-third,[91] primarily due to the advanced work on the Manned Orbiting Laboratory, a military space station approved by the secretary of defense in late 1963. Significantly, it was one of the few major "new starts" in space undertaken during the Johnson administration.

In other words, although the Cold War was still raging, policymakers were being much more selective in just what types of policy areas they were willing to see as relevant to it. As a result, there were a few programs, like spaceflight, that had benefited from that association which suddenly found themselves losing political support (and thus funding).

High-Energy Physics: In the Same Boat

It was stated earlier that the budgets for federal programs in support of basic science generally stayed level—or, in a few cases, increased slightly—during the latter part of the Johnson administration. The two major exceptions to this trend were civilian spaceflight and high-energy physics. As it happens, the experiences of the U.S. high-energy physics program are so similar to NASA's that it is worth examining in some detail.

This branch of physics examines the properties of subatomic particles (and in fact was at one time known as "particle physics"). Like spaceflight, it has very heavy capital requirements: practically all research is carried out using large, powerful (and expensive) machines, called "particle accelerators," that force atomic nuclei to collide at very high speeds. During the latter half of the 20th century, progress in the field had become quite rapid. Although this was very exciting for the scientists involved, the brisk pace of discovery put them in the unfortunate position of having to make frequent requests to Congress to fund new—meaning bigger and more costly—accelerators.

For a brief period in the late 1950s and early 1960s, that is, the first few years after *Sputnik*, this was not too much of a problem. Like their brethren in the space program, physicists had been pointing out Soviet advances in the field as a way of prodding federal officials to increase their budgets. And, again like spaceflight, it was for a time quite successful. Between 1959 and 1965, federal funding for all physics programs grew two and one half times.[92] Spending for high-energy research alone went from $53 million in 1960 to $135 million in 1964, and was expected to grow to $500 million by 1970.[93]

By the middle of the decade, however, it was clear that the political environment surrounding the field was beginning to change. Scientists had been seeking approval for a new $280 million particle accelerator, to be built somewhere in the Midwest. As before, the physics community held up the example of the USSR, which at the time had the world's largest such machine, a 70-billion electron volt (GeV) device located at Serpukhov.[94] The proposed machine would easily outclass that, operating in the range of 200 GeV. This would be followed, it was hoped, by an even larger, 600–1000 GeV, machine costing $800 million.

Much to scientists' surprise, however, policymakers were expressing some reluctance, and some were even voicing outright opposition. It is particularly interesting to note how similar these concerns were to those being expressed (at about the same time) about NASA's programs. First, there was the cost. Congressman Chet Hollifield, chairman of the Joint Committee on Atomic Energy, cautioned that "there is no end to scientific ambitions to explore, but there is an end to the public purse."[95] Meeting with a group of physicists and midwestern congressmen, President Johnson scoffed that he was "the only man in government who wants to save money and here are all these people who want to spend money."[96] In general, critics openly wondered at the practical value of large particle accelerators, given that the federal budget was under such strain.[97]

Second, many opponents of the machine—which even included some physicists—argued that its high cost diverted money from other R&D programs. Alvin Weinberg, director of the Oak Ridge National Laboratory (and who is credited with coining the term *big science*) claimed that research in particle physics did not carry over into other fields, even within physics itself. Overall, he rated the project "nil" in terms of technology development and social value, and concluded:

> Those cultures that devoted too much of their talents to monuments which had nothing to do with the real issues of human well-being have usually fallen upon bad days. . . . We must not allow ourselves, by short-sighted seeking after fragile monuments of Big Science, to be diverted from our real purpose, which is the enriching and broadening of human life.[98]

Ultimately, President Johnson killed the project in early 1964. Its demise signaled a downturn in funding for high-energy physics generally, with largely negative effects for the field overall.[99] By 1966, applicants for jobs at the American Physical Society national convention were nearly double the number of openings. Two years later, it was estimated that only 7 out of 10 physics Ph.D.s were finding jobs in the field, and that as many as one-third were making their living as postdoctoral fellows.[100] The situation led physicist Victor Weisskopf to the bitter comment:

> Here is a generation of people who studied physics under the stimulus of *Sputnik*. As kids in school they were told that there was this great national emergency and that we needed scientists. So they worked hard— it's not easy to become a physicist—and now they have maybe a wife and a child and they are out on the street and naturally they feel cheated.[101]

All Built Up and No Place to Go

A common criticism of the U.S. space program since the moon landing—one that is still heard today—is that it has never had a similar sort of long-range "goal." Space advocates often call for some sort of overarching objective (a mission to Mars is a perennial favorite) that would, in their view, serve the same function as Apollo in the 1960s.[102] Implicit in this view is the notion that the success of the lunar program stemmed, at least in part, from the dramatic, "challenge" format initiated by President Kennedy.

The analysis presented here, however, suggests that the problem was really much more basic. Although the Cold War certainly must stand as one of the most traumatic periods in U.S. (not to mention world) history, its political utility to organizations like NASA cannot be overstated. Because it was identified with the single most salient issue of the time, spaceflight became a higher-priority item than it ordinarily would have been. Moreover, this identification provided public officials strong political and personal incentives for supporting the program.

As has already been pointed out, there was not at the time (nor is there now) any other potential justification for space activity that is anywhere near as powerful politically. Thus, the loss of its Cold War definition was devastating: it struck directly at the political heart of the program—and of NASA itself. Stated bluntly (and returning to the question raised in chapter 1), by the late 1960s, there was no longer any agreement among policymakers as to what the space program—with its billions of dollars worth of technology, facilities, and personnel—was *for*.

Under the circumstances, it is hardly surprising that political leaders were unable to decide on a new "goal." This can clearly be seen in the experiences of President Nixon's Space Task Group (STG), the final attempt to develop a set of post-Apollo space objectives. Less than a month after taking office, Nixon asked his vice president, Spiro Agnew, to chair the STG, which also consisted of Secretary of Defense Melvin Laird,[103] acting NASA administrator Thomas O. Paine, Science Advisor Lee A. DuBridge, and a few other officials (including, significantly, Budget director Robert P. Mayo) as "observers." Its stated purpose was to produce "definitive recommendations on the direction the U.S. space program should take in the post-Apollo period."[104]

Unfortunately, the STG was unable to reach a "definitive" consensus. The vice president, for example, favored NASA's current course. During one of the early meetings, he called for an "Apollo for the 70s,"[105] and on the day of the *Apollo 11* launch, stated that it was his "individual feeling that we should articulate a simple, ambitious, optimistic goal of a manned flight to Mars by the end of the century."[106] The other members, however, were more skeptical, believing that, even if it were desirable from a programmatic standpoint, such an exceedingly large project stood little chance of approval.[107] Thus, like its predecessor documents during the latter part of the Johnson administration, the final STG report contains no specific recommendations, but lists a number of possible post-Apollo programs. These included a Mars mission, along with continued exploration of the moon (eventually leading to a permanent base), an orbital space station, and a reusable shuttle.[108]

Although the STG submitted its report to the president in September 1969, Nixon did not officially respond to it until the following March, largely because the dissent that marked the group's deliberations continued even after its work had been completed. In a September 25 memorandum (i.e., less than two weeks after the STG report), BOB director Mayo warned the president that endorsing any of the recommendations committed the administration to significant near-term budget increases, with the result that it could "lose effective fiscal control of the program." He recommended delaying any decision on the future direction of space policy pending a full review by the cabinet, the National Security Council, and Office of Management and Budget (OMB).[109] These reviews, and subsequent debates, lasted for several months.

Nixon ultimately ended up rejecting virtually all of the STG's candidate missions, albeit rather gently. In his March 7 statement on the future of the space program, the president stated that a flight to Mars would happen "eventually"; that a decision on a permanent space station would await the results obtained from *Skylab*; and he called for studies on the feasibility of a reusable shuttle. For the near term, he announced that the NASA budget for FY 1971 would be even less (by more than $400 million) than the year before. Space activities, he said, must "take their place within a rigorous system of national priorities," and that "many critical problems here on this planet make high priority demands on our attention and our resources."[110] Clearly, space was no longer viewed as essential to "national survival."

It is important to note that, beyond rejecting a second Apollo-like commitment, the president had relatively little to say about what NASA should do. The March 7 memo did list six "specific objectives"—continue exploring the moon, scientific investigation of other planets in the solar system, lowering the cost of space operations, extending human capability in space, expanding practical applications of space technology, and encouraging greater international cooperation—but it provided few details and almost no concrete proposals or actual projects beyond those that had already been approved (such as *Skylab*). In other words, Nixon's "specific" objectives were, in reality, rather vague.

This was, however, to become for the next several years something of a pattern in statements regarding U.S. space policy. Over the next decade, through three different presidential administrations (Democratic and Republican),[111] virtually every discussion, public and private, of NASA and American efforts in space would be characterized by three basic themes. The first was to refer to the program in the negative, that is, to describe what it was *not* doing. Nixon's 1970 statement, for example, explicitly rejected projects that required "a series of separate leaps, each requiring a massive concentration of energy and will and accomplished on a crash timetable."[112] The obvious intent here was to differentiate decision making on space policy in the 1970s from that of the previous decade.

This tendency was particularly pronounced during the presidency of Jimmy Carter. Although Carter did not appear to regard the space program as a particularly high priority (six months after taking office, the president was warned by his Science Advisor, Frank Press, that the lack of guidelines on space was placing NASA in "a difficult position" and that there were concerns over agency morale[113]), his administration did begin to develop its approach to space policy in mid-1978. At that time, Press noted that

> to focus the U.S. on a high challenge, high visibility major new space initiative does not seem feasible within any projected budget envelope or the technological opportunities on the immediate horizon.[114]

Accordingly, Carter's first major statement on space, issued in May 1978, called for a program that did not center "around a single, massive engineering feat."[115] Similarly, the Space Policy itself, announced the following October, stated that "it is neither feasible or necessary at this time to commit the United States to a high-challenge space engineering initiative comparable to Apollo."[116] Privately, Press, in a 1978 memorandum, congratulated the president on R&D policies in space that avoided "large spectaculars."[117]

Second, there were repeated declarations—to a degree that borders on the comical—that NASA and the space program should be "balanced." In 1969, DuBridge told a congressional committee:

> I see a requirement for a balanced program for the future, planned with a substantial margin of flexibility in objectives, permitting the opportunity for exploitation of new scientific findings or new capabilities that may develop.[118]

Nixon's first NASA administrator, Thomas O. Paine, echoed this sentiment during his Senate confirmation hearings, calling for "a balanced program that contains a healthy manned space flight program and also contains a strong application, planetary, aeronautics, and other programs."[119] The president himself asserted in his 1970 statement, "Our approach to space must continue to be bold—but it must also be balanced."[120]

This perspective was also prevalent throughout the Carter presidency, although there seems to have been an attempt made to find alternatives to the word *balanced* as a description. The May 1978 memo from Frank Press, for example, argued that the goal of U.S. space policy should be to "[m]aintain a vigorous, diversified, and broadly-based program of space exploration, research, and development."[121] Similarly, the October presidential directive spoke of "pluralistic objectives."[122] The final Space Policy, however, calls once again for "a balanced strategy of applications, science, and technology."[123]

Finally, virtually all official pronouncements about space began to emphasize applications, that is, using space technology to make life better "here on earth." As has already been seen, the economic and other practical benefits of space R&D—communications, weather forecasting, remote sensing, and so on—had always been included as one of the justifications of the program, but had never been considered (at least not since *Sputnik*) as its primary purpose. During the 1970s, however, "using space" began to take on a special prominence. President Nixon spoke of making a "concerted effort to see that the results of our space research are used to the maximum advantage of the human community" and of "hasten[ing] and expand[ing] the practical applications of space technology."[124] His approval of the space shuttle program (see below) was based on the belief that it would "go a long way toward delivering the rich benefits of practical space utilization and the valuable spinoffs from space efforts into the daily lives of Americans and all people."[125]

Not surprisingly, President Carter's approach to space policy strongly emphasized "applications for economic and human development."[126] The "needs of our society," stated the October 1978 presidential directive, "will set the course for future space efforts."[127] Indeed, the president left little doubt that he saw such efforts as the real purpose of the program: "the spectacular efforts to send men to the moon have been a precursor to the more practical and consistent and effective use of space technology."[128]

On the face of it, "balance" and "practicality" would seem to be laudable enough traits for any public program. Moreover, choosing not to pursue another costly space venture like Apollo was almost certainly in line with the prevailing public mood. A 1969 Harris poll, for example, found that 56 percent of the public felt that Apollo cost too much and that a $4 billion budget for NASA was too high.[129] A 1969 memorandum to President Nixon reported:

> The October 6 issue of *Newsweek* took a poll of 1,321 Americans with household incomes ranging from $5,000 to $15,000 a year. This represents 61% of the white population of the United States and is obviously the heart of your constituency. Of this group, 56% think the government should be spending less money on space exploration, and only 10% think the government should be spending more money.[130]

Unfortunately, the public statements about its objectives did very little to help NASA officials design a coherent post-Apollo space policy. *Balance*, for example, can be taken to mean virtually anything, particularly when combined with terms like *bold* (Nixon) or *vigorous* (Carter). Indeed, in 1962, while planning for Apollo was well underway, James Webb referred to NASA's programs as "well-balanced . . . in all areas."[131]

In general, one of the more striking features of this period of space history is how few specific decisions were made, other than rejecting proposed projects from NASA and elsewhere. (And even the rejections were not always the result of an actual "choice." There is no record, for example, that Lyndon Johnson ever explicitly turned down Webb's requests for a post-Apollo program; rather he delayed deciding until the only reasonably available option was to terminate production of Apollo hardware.) Far from carrying out the directives handed down from above, NASA was forced to return to administration officials time and again with new or modified proposals.

On the other hand, these types of actions—or lack thereof—make perfect sense for a policy area that no longer has any sort of issue definition attached to it. No longer identified with the Cold War, and with no other generally accepted definition to take its place, it is difficult to see how policymakers could have reached a consensus on a new mission for NASA, particularly in view of the exceedingly high cost of its programs. As seen in the previous chapter, such a "system of priorities," as it was understood at the time (e.g., a matter of "national survival"), was the impetus for Project Apollo. What was now lacking was a clear sense of how space exploration's "proper place" was to be defined.

Lock-In at NASA

To make matters worse, it is far from clear that, even if a new definition—or purpose, or mission—for U.S. space policy had been developed, NASA as an organization would have been in any position to act on it. For more than a decade, the agency's growth and development had been directed toward the primary goal of establishing U.S. preeminence in space. That goal, in turn, was based on the central premise that such preeminence could only be established through human spaceflight. As seen in previous chapters, public officials and other knowledgeable observers time and again noted that the United States was well "ahead" of the Soviet Union in space science and applications, but that it required "man in space"[132]—and, later, "man on the moon"[133]—to demonstrate "true" dominance. Acting on this premise, NASA had, with presidential and congressional approval and encouragement (and even, at times, outright prodding) invested substantial sums of public money to develop the facilities, technology, and infrastructure intended to achieve this end.

Chapter 2 described the phenomenon of "technical lock-in," in which an organization has become so heavily invested in a particular set of technologies or in one technological approach that it cannot, either for financial reasons or because of organizational inertia, deviate from it, even when doing so might be beneficial in the long run. This phenomenon seems to describe NASA's position by the 1970s. The types of "applications" that were then under discussion—communications, remote sensing, and so on—seldom required human presence in space. Indeed, as has been noted earlier, critics of Mercury, Gemini, and Apollo (including the Apollo Applications Project) frequently maintained that the cost (and the risk) involved in human spaceflight could not be justified by its economic or scientific returns. Moreover, it could be argued (as many did) that the very large proportion of NASA's budget devoted to the manned program did not fit into any reasonable notion of "balance." Finally, the explicit rejection of large-scale space endeavors, such as the Mars mission or a permanent space station, would seem to preclude any new ventures involving humans in space. In short, a case could be made that a "new," post-Apollo NASA would look very different, with little, if any role for the astronaut corps.

This was a course that many at NASA, and the space community generally, could reasonably have been expected to resist. To begin with, most (although not necessarily all) space enthusiasts were personally committed to keeping the human spaceflight portion of the program. They regarded (and many still regard) it as an indispensable part of the program, part of our "destiny" as a species. Second, such a retrenchment involved much more than simply not writing any more checks. Dismantling the vast infrastructure devoted to astronaut operations would have required a massive effort (and almost certainly great expense). Finally, by 1970, human spaceflight represented more than half of NASA's budget (and was widely seen as its primary source of visibility and prestige). To many involved with the space program, eliminating, or even downgrading this activity would have amounted to an unacceptable "gutting" the agency.[134]

As it happened, however, at least part of NASA officials' feelings were shared by some in the White House. In a memo to President Nixon, Caspar Weinberger (who was then serving as a deputy director of the Office of Management and Budget) contended that such a cutback "would be confirming, in some respects, a belief that I fear is gaining credence at home and abroad: that our best days are behind us, that we are turning inward."[135] For his part, Nixon was becoming concerned—particularly after the cancellation of the Supersonic Transport—that too much downsizing of the program might hurt the Republican Party (and perhaps him personally) in states like California that were heavily dependent on the aerospace industry.[136] In addition, he did not wish to go down in history as the president who ended the era of man in space.[137] Evidently, a sort of "political lock-in" was also operating within the executive branch.

The Space Shuttle: An Undefined Technology

This, then, was the policy environment that gave rise to what is officially known as the Space Transportation System (STS).[138] Destined to become one of the largest and most controversial programs in NASA history (see chapters 7 and 8), it would represent, to its many critics, a clear case of political manipulation and technological ineptitude, a symbol of everything that had "gone wrong" with the agency. Clearly, as will be seen in the following chapters, the space shuttle fell far short of NASA's rather extravagant promises of the 1970s. Its technical and economic shortcomings are also all too obvious. Still, it is possible to argue that STS was very much a product of its time. Viewed in its larger political and historical context, it is a near-perfect example of a technology designed by an agency with no clear mission: the means for implementing an undefined policy.

Means, in fact, is the critical term here. The shuttle was originally conceived as simply one of the supporting elements of NASA's post-Apollo program, providing service to a permanent space station (proposed in the STG report). By 1970, however, it had become NASA's sole rallying point. Having failed to gain approval for any of its other proposed programs, STS was the agency's last hope for maintaining the human spaceflight program once the Apollo flights were completed. Unfortunately, since the other missions it was intended to support had been rejected, the shuttle would have to be justified to the president and Congress on its own merits.

Accordingly, NASA officials sought to package the program in a way that fit into the new ethic surrounding U.S. space policy, which meant emphasizing its economic benefits. A reusable spacecraft, they argued, would dramatically lower launch costs (as low as $100 per pound[139]), thereby providing "routine access to space." This, in turn, would open up the space environment to more users than ever before: commercial opportunities and scientific research, for example, would be greatly expanded. In addition, since its costs were to be below that of expendable launch vehicles, the shuttle could be used to launch satellites (including those of the Department of Defense, a key selling point) and deep-space probes, as well as to repair, maintain, and even return objects from earth orbit.

Thus, from the start, STS was depicted not as a "goal" in and of itself, but rather as a set of abilities, the means to a variety of other ends. Indeed, James Fletcher (appointed by Nixon to be NASA administrator in 1971), who disliked "highlighting the cost-benefit argument," stated that the shuttle's "most important justification" was the "entirely new *capability* for working routinely and quickly in space."[140] This particular way of visualizing the program, it should be noted, extended to NASA itself. From the agency's point of view, the success of STS would pave the way for other, more ambitious projects: a permanent space station, lunar bases, and so forth.

Of course, justifying a multibillion dollar program on the basis of what it could do, without clearly stated, specific objectives, was a somewhat risky strategy. Many critics, for example, attacked STS as "lacking a mission." Such a strategy was, however, completely rational (which is not to say necessarily justifiable) when seen in the context of the post-Apollo political environment. During the 1960s, the Cold War had provided NASA with a constituency large enough to support its major programs. By 1970, there was no longer any single objective (or even a set of objectives) that would generate the same level of political support. Thus, to secure approval for the shuttle (or any other large-scale project), NASA was forced to piece together a coalition of supporters by promising to provide a (very) broadly based service.[141]

The various technical and budgetary travails that accompanied the shuttle's development have been widely discussed elsewhere.[142] Although NASA had initially estimated that the R&D for STS would cost $15 billion (later lowered to $10 billion), it ultimately had to settle for $5.5 billion. This, in turn, required a substantial redesign, resulting in a system that was only partially reusable (the external fuel tank would be discarded on each flight). The agency also encountered some difficulties in developing a design that would fully satisfy all of its potential customers (particularly DOD), and struggled (ultimately unsuccessfully) with making the overall system truly cost-effective. As a result, the program fell far behind schedule. It also ran well over its budget, forcing the agency to divert funds from its other programs (most notably space science[143]). All of this turmoil might well have been justified if STS had performed up to the standards originally set by NASA (there is an old saying in the theater that "nothing is as cheap as a hit, no matter what it costs"), but despite the initial excitement following the first shuttle launch in 1981 (see next chapter), the shuttle ultimately fell far short of those expectations.

Conclusion: Ripe for a Change

A document prepared for the Carter–Mondale presidential transition team noted that "NASA has more difficulty than most agencies in describing national goals in a way that its programs relate to them" and that "[m]uch apprehension and uneasiness about the NASA budget would disappear if the civilian space program, like its military counterpart, had clear objectives related to national goals."[144] Although this assessment was, on the face of it, essentially correct, the analysis presented here suggests that this was not a problem of NASA's making. For much of the previous decade, the agency had been closely identified with one of the nation's most important "goals." Moreover, that goal had been articulated, not by NASA, but by President Kennedy, President Johnson, and the U.S. Congress. It can therefore be argued that the agency's "difficulty" resided less in its own shortcomings than in the unwillingness of elected officials (beginning, ironically, with Johnson) to identify the policy problem(s) NASA was expected to address.

In short, what the space program needed—badly—was a new definition, a connection to an ongoing and compelling public issue. Although some policymakers still expressed mild concerns about the USSR in space,[145] few people—in and out of government—would have accepted once again defining NASA in Cold War–based terms. This new definition would have to be drawn from some other policy area. As the next chapter will show, the results of the 1980 election began to move NASA and the space program in precisely this direction.

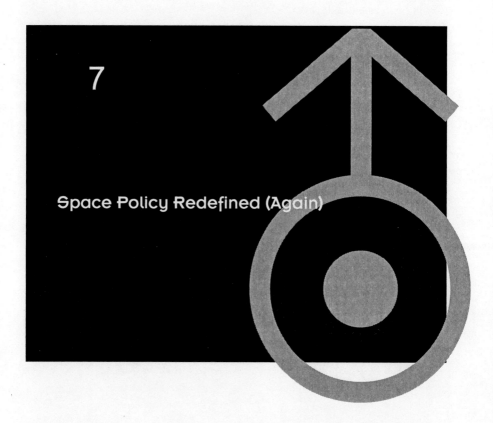

7

Space Policy Redefined (Again)

During the latter part of the 1970s, U.S. space advocates had predicted that the following decade would see spaceflight becoming "routine," with NASA's new space shuttle making as many as 40 or 50 flights per year. Although the STS itself came nowhere near achieving this goal (the actual number of flights in 1989, for example, was six), it can be argued that in another, perhaps more meaningful, way space technology did make major strides toward a sort of "routinization."

As this chapter will show, for the first time since the 1960s, space technology played a significant role in a number of "conventional" areas of government policymaking during the 1980s. In other words, spaceflight was transformed—that is, redefined—from an end in itself (or as a path to "future benefits") into a mechanism for achieving other policy ends. It is therefore quite likely that the 1980s will be remembered as a major turning point in the history of space exploration, possibly as important a decade as the 1960s.

This was due in large part to a convergence of two very different, but ultimately intertwined, series of events. First, space technology itself had matured to the point where, while not exactly as "routine" as aeronautic or marine transport, it had become more fully integrated into the American civilian economy. Put another way, many aspects of space operations now no longer needed to be performed exclusively by NASA (or even by government, for that matter). Moreover, it appeared that a number of space applications, such as rocket launching, remote sensing, and perhaps even space-based materials processing were on the verge of following communication satellites (which had been commercially viable for many years) into the marketplace.

Second, the U.S. political scene experienced a profound change. The election of Ronald Reagan to the presidency in 1980 (and, to a lesser extent, the takeover of the Senate by the Republican Party) brought to power in Washington a new group of policymakers who sought to remake virtually every federal policy and program. In particular, these men and women had very definite ideas about using space technologies to attain these ends. The result was, by the end of the Reagan presidency, a very different approach to space policy and a lot of questions for NASA, many of which have still not been answered.

Ronald Reagan, Conservatism, and R&D

There is a strong tendency among political scientists and historians to portray the Reagan presidency as being uniquely monolithic in its outlook. Some, for example, describe his administration as being the most "ideological" since Franklin Roosevelt's,[1] or "the most explicitly pro-business . . . since the 1920s."[2] In fact, Reagan's government, like those examined in previous chapters, had its fair share of differences over goals, priorities, and approaches to governing, even among those personally loyal to the president. Since (as the discussion below and in chapter 8 will show) these divisions were to be a factor in the administration's approach to space policy, it will be useful to explore them—and their sources—in some detail.

Internal Policy Differences

To begin with, it is a great oversimplification to refer to the Reagan presidency simply as "conservative" with no further elaboration. Like most political philosophies, modern conservatism comes in a number of varieties.[3] The supply-side theories that guided much of Reagan's economic policies, for example, were not universally accepted by "traditional" conservative economists.[4] Similarly, the so-called religious right does not have the same set of priorities or approach to policymaking as do free market libertarians.

Second, the Republican Party during the 1980s, although certainly "conservative" in its overall makeup, represented a coalition of somewhat diverse—and, in some cases, contradictory—interests,[5] including at various times traditional Republicans, social and religious conservatives, and the group of southern whites and northern blue-collar workers who came to be known as "Reagan Democrats."[6] In addition, one of the more significant—and least remarked on—additions to the Reagan coalition was an unusually unified business community. Although often seen as exclusively Republican in their sympathies, business interests have, in most postwar elections, given support to candidates from both parties (so that they will continue to have access regardless of the outcome). In 1980, however, virtually all business interests lined up behind one candidate: Ronald Reagan.[7] While this coalition was highly successful from an electoral standpoint, winning two national elections by wide margins, it did run into difficulties when it came to formulating and implementing policy, particularly—as will be seen—with regard to the space program.[8]

Finally, even a broad-based consensus over a matter of principle can, on occasion, break down over questions of how that principle should operate in practice, or how it applies in a specific situation. This tendency is sometimes exacerbated by the propensity of individuals occupying different policymaking positions to view problems according to their particular area of responsibility (i.e., "where you stand depends upon where you sit"). It is generally the case, for example, that a budget director, regardless of administration or party affiliation, always comes across as more fiscally conservative, and more suspicious of high-cost programs, than other cabinet members or agency heads. This comes, in large part, from that individual's position as "guardian of the public purse."[9] Thus, it should come as no surprise when the secretaries of state or defense see an issue differently than the director of OMB or the administrator of NASA, even if all adhere to the same general philosophy and are serving the same president.

Man and Administration

Second, it is always important to make distinctions between the views, policies, and proposals of various administration officials and those of Reagan himself. Most personal descriptions of the president portray him as a man more focused on a larger vision of America than with the policy details needed to implement those visions. Thus, he was at times predisposed to make statements that were at variance with established fact, or even with (in the view of some of his advisors) conservative principles. As one reporter noted in a famous *New Yorker* article:

> Reagan understands the importance of having a vision and stating it forcefully, and knows that this can be far more powerful than facts. People who intrude with facts are "doomsayers" and "handwringers" who should be ignored.[10]

Or as a group of *Newsweek* reporters observed, "[H]e sketched his outline of a vision and left others to color it in. He resisted detail and endured conferences with jokes and stifled yawns."[11]

Combined with the multiplicity of views outlined above, the tendency of the president to operate at such a high level of generality meant that, at its worst, the administration was "unable to speak with a single voice" on important issues.[12] This could be particularly troublesome in a policy area such as space, which readily lent itself to Reagan's lofty and sweeping rhetoric, but which at the same time required an unusual amount of careful and detailed planning (not to mention the fact that is also one of the larger and more expensive of government programs).

In short, the Reagan presidency, for a variety of reasons, was characterized by a multiplicity of political and ideological goals. Not only did different members of the administration sometimes pursue vastly different policies, but the president himself would on occasion take a position or embrace a goal that would surprise—

and even outrage—aides and supporters alike.[13] This sort of "fractured" policymaking shows up from time to time in decisions on the space program, leading at least one author to conclude that the administration really had no space policy at all.[14]

Grand Themes

Its internal differences notwithstanding, there were a number of basic principles that remained relatively constant throughout the Reagan presidency and that did, to varying degrees, influence its approach to the space program. First and foremost among these was a belief in rolling back the size and power of the federal government. In his first Inaugural Address, the president declared that "government is not the solution to our problems; government is the problem."[15] He would often describe the federal bureaucracy as "overgrown and overweight," an "automatic spending machine" that employs "thousands upon thousands of bureaucrats, researchers, planners, managers, and professional advocates who earn their living from the great growth industry of government."[16]

The administration moved to scale back the size of government on a variety of fronts. First, it was part of the justification behind the large tax cuts enacted in the first few months of Reagan's first term. These were, of course, a reflection of the widely held view among Republicans that American taxes were too high overall. They also were intended (consistent with the president's belief in supply-side economics) to provide a fiscal stimulus and promote economic growth. In addition, however, the reduction in government revenues (and resulting record budget deficits) were also aimed at inhibiting, if not preventing outright, new public programs. Within a few years, scarcity of funding led to all proposals for new spending receiving a much higher level of scrutiny than ever before. In effect, Reagan's tax policies redefined virtually all major public initiatives as "budget issues,"[17] and (with some important exceptions, described below) made it far more difficult to get approval for any major new programs.

Second, Reagan actively sought to reduce the number and scope of federal regulations governing business activity. Critics had long argued that "needless" regulatory activity was costly for consumers. Murray Wiedenbaum, Reagan's first chairman of his Council of Economic Advisors, once claimed that compliance with all federal regulations cost Americans more than $100 billion (or $500 per capita) a year in larger, "hidden" prices.[18] In addition, it was said, the resources, time, and paperwork required to deal with the government's regulatory demands had cut into business' research and development budgets, resulting in U.S. companies becoming less innovative.[19] Although significant changes in regulatory policy had taken place in the Carter administration (most notably in such areas as trucking and commercial airlines), deregulation was to become one of the central themes of the Reagan presidency.

Third, the administration engaged in a major effort at privatizing a large number of federal programs that involved the production or distribution of goods or services. The term *privatization* was coined by the libertarian Reason Foundation in 1976,[20] although the doctrine that:

[t]he federal government will not start or carry on any commercial activity to provide a service or a product . . . if such product or service can be procured from private enterprise through ordinary business channels

actually dates back to the Eisenhower administration.[21] The fundamental premise behind privatization is that, except for public goods (like national defense), the market forces that shape private sector behavior are far more likely to produce optimum results (higher quality, lower cost, etc.) than is the "command-and-control" method of operating inherent in a government bureau.[22]

There are two primary approaches to privatization.[23] In so-called load shedding, the government sells (or transfers in some other fashion) a publicly owned asset to a private interest. In 1987, for example, the Department of Transportation sold Conrail, a freight railroad that the government had taken over from the bankrupt Penn Central 10 years earlier, for $1.6 billion public stock offering. The second major privatization arrangement is contracting, in which a good or service utilized by a government agency is actually provided by a private company under a fixed-term contract with that agency. Since the government always has the option (in theory) of switching to another firm when a contract period is up, the company that provides the good or service is said to have a strong incentive to operate as efficiently as possible.[24] Contracting has become increasingly popular in recent years, at all levels of government. To take only one case in point, a number of federal agencies, such as the National Science Foundation, have recently moved from government-owned buildings into spaces leased from private office complexes.

In Ronald Reagan, privatization was to find its most ardent champion to date, at least rhetorically. Along with Conrail, the administration sought, at one time or another, to privatize the National Institutes of Health's intramural research activities (although not its grant-making function),[25] the Amtrak passenger rail system, the Overseas Private Investment Corporation, a number of insurance and loan programs, several oil fields owned by the Department of Energy, and the National Weather Service and the Landsat remote-sensing satellite system (see below).[26] Most of these efforts were rejected by Congress.

As noted earlier, Reagan was held to be more "pro-business" than most of his predecessors. This did not necessarily mean, however, that government would automatically come to the aid of private firms. The administration's abiding faith in the free market system also meant that, in theory at any rate, the private sector was to sink or swim on its own. Thus, the 1980s saw the beginning of the still-raging debate over "industrial policy," programs intended to aid selected firms or "strategic" economic sectors.[27] For the most part, conservative Republicans usually reject this approach to economic development as just another name for government "planning"[28] that replaces "the test of the marketplace with raw political power"[29] and puts bureaucrats in the position of "picking winners."[30]

The other "grand theme" of the Reagan presidency was a strong commitment to national defense, combined with a fierce anticommunism. Characterizing the 1970s as a "period of neglect," the administration set out to "make America strong again." Toward that end, it proposed the largest and fastest buildup of defense-related budget authority in peacetime history. Between 1980 and 1985, the DOD budget grew by nearly 53 percent (by $1.5 *trillion* over the 5-year period), with spending for procurement more than doubling (in real terms).[31]

Although it might appear (and reportedly did appear to some of his domestic advisors) that the exceptionally rapid and expensive buildup of the military was at odds with his other frequently expressed determination to "shrink" government, Reagan evidently saw no such conflict. To him, keeping the armed forces strong and well-supplied was the "first duty" of government, the primary means of protecting the freedom and individual liberty that he hoped to achieve through tax cuts, deregulation, and privatization. Thus, as one observer put it, he kept defense expenditures "in a different part of his brain" from the budget for domestic programs.[32]

Reagan's personal attitude toward the USSR, which was generally shared by everyone in his government, was clearly expressed in January 1981, at his first presidential press conference. Responding to a question about U.S.–Soviet détente, the new president said:

> [The Soviets] have openly and publicly declared that the only morality they recognize is what will further their cause, meaning that they reserve unto themselves the right to commit any crime, to lie, to cheat, in order to attain [world revolution].... [W]e operate on a different set of standards. I think when you do business with them, even at a détente, you keep that in mind.[33]

Not surprisingly, during the first three years of Reagan's first term, U.S. relations with the Soviet Union reached their lowest level in many years. At one point, Georgi Arbatov, the leading Soviet expert on American politics and culture, even accused Reagan of acting like Adolf Hitler.[34]

Because of his distrust of the USSR, the president tended to regard arms control agreements, such as the Strategic Arms Limitation Treaties negotiated by his predecessors, with a high degree of skepticism. In a 1981 speech at West Point (where he referred to the Soviet Union as an "evil force"), he declared that

> no nation that placed its faith in parchment or paper, while at the same time giving up its protective hardware, ever lasted long enough to write many pages of history.[35]

This need for "protective hardware" referred not only to weaponry, but also to his insistence that any arms control agreement with the USSR must contain some means of effective verification. This, in turn, was the basis of his aphorism of "trust but verify," a phrase he repeated so often that Soviet president Mikhail Gorbachev would hold his hands over his ears every time he heard it.[36]

Reductions in spending, deregulation, privatization, national defense, dealing with the Soviet Union over nuclear weapons: to a remarkable degree, space policy and space technology (as it had developed by the early 1980s) intersected in one way or another with each and every one of these issue areas. Combined with Reagan's personal feelings about the "space frontier" (see below), the stage was set for a number of major developments in the U.S. space program—except for one last item.

What Is Conservative Science Policy?

The only area of federal R&D policy in which the application of conservative political and economic principles appears to be relatively straightforward is when it is in support of national defense. Obviously, if raising and maintaining an army is a (if not the) primary duty of government, it follows that research and development activities carried out on its behalf must be considered a public good as well. It is therefore not surprising that the budget for defense-related research and development rose rapidly throughout the Reagan years.[37]

With regard to civilian science and technology, however, the situation is less clear. The dominant conservative position is that, in most cases, the best way to stimulate scientific advancement and the creation of new technologies is "to reward private inventors, entrepreneurs, and investors."[38] Government support, if it is to be provided at all, should be limited to establishing "a supportive climate for private initiative and individual enterprise"[39] or, if more direct means are needed, by providing tax incentives.[40]

Most conservative commentators acknowledge, however, that there are cases, most notably basic research (i.e., investigations into fundamental scientific questions[41]), where—despite its evident value—the costs are so high, the (financial) risks are so great, and the rewards (in the form of commercial profit) are so distant that private industry cannot (or will not) make the necessary investments no matter how favorable the business climate or generous the government's tax breaks. Thus, the Reagan administration accepted as a general principle the notion that the federal government has a broad responsibility to support "basic research across all scientific discipline,"[42] with a particular emphasis on "high-cost, high-risk, long-term research."[43] By 1985, 42 percent of all federal nondefense R&D funding went to basic research, compared to 29 percent in 1981.[44] Moreover, by the end of Reagan's eight years in office, government funding for basic research had nearly doubled (after initially being cut; see below), from $5.9 billion in 1980 to $11.7 billion.[45] William Carey, executive director of the American Association for the Advancement of Science, went so far as to declare that scientists across the country must be "pinching themselves" over their good fortune.[46]

In some respects, it is a rather simple rule of thumb: government funding is appropriate in the case of vital scientific research that, for reasons of cost or profitability, is beyond the reach of the private sector. Such reasoning, however, is not necessarily relevant to the processes of applied research (investigation motivated by a direct, recognized need[47]), or technological development (the direct production of materials, designs, processes, or prototypes[48]). Unlike basic research, these activities seem, at first glance, to be more appropriate for private industry, especially since they are commonly the last step before actual commercialization.

When viewed in this way, government involvement in this type of R&D begins to resemble direct aid to business, the same sort of "industrial policy" and "picking winners" that conservatives generally decry.[49]

Clearly, Democratic presidents have had little problem with this idea. The Carter administration was quite fond of so-called demonstration projects—particularly those related to energy production, such as solar, nuclear fusion, oil shale, and coal gasification—intended to demonstrate technical feasibility or commercial potential.[50] Indeed, by FY 1980, the last Carter budget, the federal government accounted for nearly half of all U.S. spending on applied research, almost as much as industry, universities, and nonprofit sources combined.[51] President Bill Clinton also instituted a number of controversial (particularly among Republicans) programs that provide direct federal aid to private firms for the pursuit of high-risk, "precompetitive" technological development (see next chapter).[52]

Consistent with conservative principles, Reagan officials generally opposed such policies, at least up to a point. Certainly, the free market rhetoric was there. The president's 1984 *Economic Report*, for example, declared:

> Some industrial policy proponents advocate government aid to "linkage" [i.e., strategic] industries. . . . Steel and semiconductors are often cited as examples of "linkage" industries. However if such an industry is vital, then the industries that rely on it will demand its output.[53]

OMB director David Stockman, characterizing Carter's energy R&D policies as unwarranted handouts to big oil companies, called industrial experimentation "precisely the kind of thing that Adam Smith invented the free market to accomplish."[54] Reagan's first science advisor, George Keyworth, in announcing the elimination of programs like solar energy research, stated that the administration intended to create a "clear-cut . . . distinction between what should be public sector and what should be private sector responsibilities."[55] Thus, most authors view the Reagan administration as far less likely to approve of "interventionist" technology policies than those of the Carter, Clinton, or even the Bush administrations.[56]

As would be expected, most of the Carter energy projects were either substantially downgraded or canceled outright when Reagan took office (with one important exception; see below).[57] There was even—for a brief time—an effort to roll back the federal government's long-standing commitment (i.e., back to the founding of the National Advisory Committee on Aeronautics in 1918) to aeronautical R&D. Believing that the aircraft industry was wealthy enough to fund such research on its own, Keyworth convened a special panel of the White House Science Council in 1982 to consider phasing out those programs, then part of NASA. Noting, however, that such areas as aerodynamics, safety, and common-use technologies were receiving "insufficient" attention from the private sector, the panel recommended that such funding be maintained, if not increased.[58] Overall, by the end of the president's second term, federal spending for applied research had fallen to less than one-third of the national total.[59]

Nevertheless, this reluctance to support "industrial" research and development was far from absolute. To begin with, although government funding for applied R&D grew more slowly during the 1980s than it had in previous decades, by 1988 it was still more than $10.5 billion, an increase of about 40 percent since 1980.[60] In addition, Reagan officials did maintain at least one Carter-era energy program. Even as the budgets for research into fossil fuels, energy conservation, and solar power were experiencing significant budget cuts (if they were not wiped out altogether), spending on nuclear fusion research remained relatively intact (which is to say that it was not cut as heavily as were the others).[61]

The administration—and sometimes the president personally—also were not above pursuing large-scale, high-tech demonstration projects of their own. Notwithstanding the fact that the 1980s were a time of great budget stringency, the Reagan presidency at one point or another gave its support to such costly—and controversial, even among Reagan officials[62]—programs as the Clinch River Breeder Reactor,[63] the Superconducting Super Collider,[64] and NASA's space station.

Finally, pushed by Congress, the R&D community, and business leaders, the Reagan administration did, toward the end of the second term, begin to enter into some cooperative arrangements with private industry on selected technologies. The most well-known of these is the Sematech, a public–private research consortium established in 1987 to advance semiconductor technology.[65] Other initiatives provided assistance to the fields of biotechnology, superconductivity, and high-speed computing.[66] In fact, by the end of the Reagan presidency, Robert Reich (who would later become President Clinton's first secretary of labor) was claiming that "[r]arely has an Administration sought more actively to encourage specific industries and technologies."[67]

To some extent, this allegedly unprecedented level of federal involvement in industrial R&D could be defended on grounds similar to those relating to basic science. According to the administration, support could be granted to "technologies requiring a longer period of initial development."[68] This would, for example, justify continued government funding of fusion research, given that the field is considered to be still on the "cutting edge" of plasma physics, and thus not yet viable enough to attract a sufficient level of private investment.[69]

There was, however, an additional element to these programs beyond a simple correction of a presumed market failure. Reagan himself often invoked a political theme that had become a major public concern in the 1980s: U.S. competitiveness in the international marketplace. His 1983 State of the Union Message, for example, notes:

> We Americans are still the technological leader in most fields. We need to keep that edge. . . . To many of us now, computers, silicon chips, data processing, cybernetics, and all of the other innovations of the dawning high technology age are as mystifying as the workings of the combustion engine must have been when that first Model T rattled down Main Street, USA. But as surely as America's pioneer spirit made us the industrial giant of the 20th century, the same pioneer

spirit is opening up on another vast front of opportunity, the frontier of high technology.

In conquering the frontier we cannot write off our traditional industries, but we must develop the skills and industries that will make us a pioneer of tomorrow. This administration is committed to keeping America the technological leader of the world now and into the 21st century.[70]

Science Advisor Keyworth, among others,[71] echoed this idea, reminding scientists that the administration's generosity in R&D funding was not an "entitlement," but a reflection of the fact that "our leadership in the international marketplace is at stake."[72]

Indeed, the relationship between R&D and American "economic growth," "jobs," and so on would come to be heard of so much that Reich would remark in 1988 that "[n]ever has an Administration so often justified its interventions by appeals to American competitiveness."[73] The end result, however, is that administration research and development policies, including those related to space, arguably did, on occasion, stray into the realm of industrial policy, despite the seeming contradiction with conservative economic principles.

Reagan's (and others') repeated references to American "leadership" are quite significant. They are, after all, somewhat reminiscent of the themes struck by Presidents Kennedy and Johnson in calling for an expanded space program. Unlike that earlier era, however, in which "leadership" was loosely construed as "demonstrating" U.S. economic, political, and technical superiority, the Reagan administration had some very clear ideas on how it wanted America to "lead," and how space policy was to contribute.

Prelude: 1980–1981

Ironically, the space program—as well as science policy in general—was not at first a terribly high priority for the new administration. During the 1980 presidential campaign, candidate Reagan had relatively little to say about either topic, other than as part of a general criticism of the Carter administration (e.g., for not "responding to Soviet moves to establish military superiority in space"[74]). The primary transition document (the self-proclaimed "'Bible' of the Reagan Administration"), prepared by the Heritage Foundation, also said very little about federal R&D—a chapter on technology written by the famous physicist Edward Teller was confined solely to discussion of military technology.[75]

What little was known was not particularly encouraging to the science and engineering community. In fact, the scientific community initially regarded Reagan with some skepticism. Many scientists' eyebrows were raised, for example, after he had questioned the theory of evolution during a speech before a group of ministers.[76] During the transition period following the election, science professionals were alarmed to learn that a former congressional aide who had criticized the National Science Foundation's (NSF) education programs during the 1970s for promoting "secular humanism," a "liberal, anti-religious" ideology in public schools was on the president-elect's transition team, charged with formulating new policies for that agency.[77] Bruce Murray, director of NASA's Jet Propulsion Laboratory, recalled examining a list of Reagan's close friends and advisors, and was dismayed at not being able to find a single individual with any "personal competence" in science, engineering, or technology (he then proceeded, however, to describe Reagan's "political instincts about man in space" as better than any president since Kennedy).[78]

Even after entering office, the personal views of the new president were to remain a bit of a mystery for some time. As of April 1981, the eve of the first shuttle launch, Reagan had not yet appointed either a NASA administrator or a science advisor, and no one at the White House had been assigned (at least officially) to deal with space-related issues.[79] This was, some felt, a deliberate strategy to allow the president to distance himself from NASA in case the *Columbia* mission failed.[80]

The difficulties in selecting a director for the Office of Science and Technology Policy (OSTP) (i.e., the science advisor) seemed to confirm the scientists' worst fears about the kind of administration they were dealing with. Rather than report directly to the president, as previous OSTP chairs (and their predecessors) had done, the Reagan White House placed the science advisor under Domestic Policy Advisor Edwin Meese. White House officials were said to be cool to the idea of a science advisor anyway, believing that the post had degenerated into nothing more than another mechanism by which scientists pleaded for more funds.[81] Reportedly, a large number of noted scientists turned down the position specifically because of the lack of presidential access.[82]

While all of this was happening, the first Reagan budget was made public. It called for the elimination of NSF's education programs along with sharp reductions (on the order of 50 percent) in its social science research budget and funding for biomedical research.[83] In addition, the NASA budget was cut by about 9 percent, with the space sciences program alone absorbing a reduction of over 22 percent (approximately $170 million). Budget director Stockman had wanted NASA to cancel its *Galileo* Jupiter probe,[84] but the agency instead chose to terminate the U.S. portion of the International Solar-Polar Mission, a joint project with the European Space Agency (ESA) to explore the polar regions of the sun. Although the European half of the mission did go forward (renamed *Ulysses*), the American cancellation had a chilling effect on NASA–ESA relations for some time after.[85] There were budget increases for some basic sciences, but these were not necessarily coordinated in any fashion, resulting in some odd decisions. Space Astrophysics, for example, received a healthy increase, while spending on planetary astronomy was cut. According to most outside observers, these apparently inconsistent policies were largely due to the lack of a central organizing figure representing the sciences, the role usually played by the science advisor.[86] In short, for the first several months of his presidency, Ronald Reagan did not appear to even have a science policy of any sort, let alone a plan for the U.S. space program.

Congress did involve itself to a small degree in filling in the void. It rejected, for example, some of the president's proposed cuts in NSF education and social science programs, and smoothed out some of the maldistribution within the agency's research budget. Individual members also lobbied for the continuation of *Galileo*. On the other hand, it did not act to save International Solar-Polar Mission, nor did it restore NASA's space science funding. It must be remembered that the first Reagan budget also called for deep cuts in a large number of other government programs, and most representatives and senators were fighting the White House on issues like public housing and welfare.[87]

All of this began to change—albeit slowly—as the year progressed. During the summer of 1981, the Senate confirmed presidential appointments for science advisor as well as for administrator and deputy administrator of NASA. For science advisor, Reagan nominated George A. Keyworth, a 41-year-old physicist from Los Alamos National Laboratory (and a protégé of Edward Teller, a fact that would become significant during the debate over the Strategic Defense Initiative[88]). Keyworth was not well-known outside of Los Alamos, and had relatively little experience in Washington politics. Moreover, he did not start out with much political capital among scientists and engineers, due to the poor first impressions conveyed by the administration's budget proposals and the perception of his lowly status within the White House hierarchy. To make matters worse, Keyworth did not appoint a White House Science Council until the following February, nor did he ever have sufficient staff support (reportedly, he had to borrow staff from other agencies and organizations).[89] Even so, Keyworth was ultimately able to garner a reputation as an effective spokesman for the administration.[90]

The new NASA administrator was James M. Beggs, an aerospace executive with General Dynamics who had headed the agency's research programs during the 1960s. He had moved to the Department of Transportation in 1969, due in large part to the looming reductions in NASA's budget. Before accepting the administrator position, Beggs sought assurances that the president did not intend to "assassinate" the agency.[91] Meeting with Reagan in March 1981, he was told that "although he didn't know much about it," the president felt that the space program "was something that the government ought to do." It did not appear to Beggs that Reagan had any specific plans for NASA at the point (neither did Beggs).[92]

Of course, one of the most important events of this period was not political, but technical. On April 12, 1981 the space shuttle *Columbia*—the world's first reusable spacecraft—made a near-flawless first test flight, followed by equally successful missions in November and two more through mid-1982. For nearly six years, while the STS had been encountering numerous political, budgetary, and technical problems (see last chapter), the USSR had had human space flight entirely to itself. Between the 1975 *Apollo–Soyuz* mission and STS-1, the Soviets had flown 31 cosmonauts (some more than once) on 22 missions. They had orbited two space stations (*Salyut 5* and *Salyut 6*), which had housed crews for missions as long as 184 days. With just one flight, however, the United States appeared to have leapfrogged completely over the Soviet program, an apparently stunning display of American technological capability.

Although the initial appearance of success would prove to be premature, the general feeling within the White House after *Columbia* was that anything was possible.[93] As members of the administration sat down in July 1981 to begin formulating their approach to space policy, it seems clear that the shuttle's presumed capabilities had an impact on their discussion. A July 17 National Security Council memo, for example, requests that the use of STS in antisatellite (see below) and even in "active military operations" be included as part of the national space policy, and that the committee drafting the policy consider "turning the Shuttle over to the U.S. private sector."[94] In short, STS appeared to provide the Reagan White House with the final ingredient—the requisite technology—that it needed to integrate the U.S. space program into its larger political and economic goals.

Step One: Redefinition

One of George Keyworth's first assignments as the new science advisor was to review the direction of the entire U.S. space program. Keyworth and others within the administration worked at this task for almost a year, and the final product was issued as National Security Decision Directive (NSDD) 42, dated July 4, 1982 (timed to coincide with the shuttle's final "test" mission and the beginning of its "operational" status).[95] Although there were to be a number of "official" statements on space policy associated with the Reagan presidency, they all share the same underlying philosophy as this document.

Stated simply, NSDD 42 marks nothing less than the beginning of the *redefinition* of U.S. space policy. For the first time since *Sputnik*, government officials revisited—and substantially revised—the objectives, values, and even the fundamental meaning of the space program. Obviously, it is worth examining the document in some detail.

In its introductory section, the directive described the "basic goals" of U.S. space policy. Some of these were the usual declarations dating back to the 1950s, such as the need to "maintain space leadership" and to "cooperate with other nations in maintaining the freedom of space for all activities that enhance the security and welfare of all mankind."[96] Other parts of the list, however, were quite different.

Defense

To begin with, the first "basic goal" was to "strengthen the security of the United States." Even at the height of Cold War–driven space policy (see chapter 4), the Eisenhower-era *Introduction to Outer Space*, defense was only listed second (after "the compelling urge of man to explore and discover"),[97] and the nation's first official space policy of 1958 speaks of scientific, military, and political purposes, in that order.[98] It should also be understood that the term *security*, as used here, was meant not in the broader—some might say overly vague—Cold War sense used by politicians during the late 1950s and early 1960s, but refers specifically to *military* space activities.

Indeed, military applications were highlighted throughout the directive. Following the "basic goals," it listed a set of "basic principles" that were to govern U.S. space policy. The first of these began in a (once again) fairly standard fashion, declaring that "the United States is committed to the exploration and use of space by all nations for peaceful purposes and for the benefit of mankind."[99] Unlike other such documents, however, there followed a second, clarifying statement that "*'[p]eaceful purposes' allow activities in pursuit of national security goals.*"[100] In addition, another principle stated that the "United States will pursue activities in space in support of its right to self-defense."[101]

All of this was necessary, according to another administration document, because "[t]he Soviet Union has initiated a major campaign to capture the 'high ground' of space."[102] Clearly, then, the White House saw space technology as another component its expansion of the nation's conventional and nuclear forces generally, as well as its more confrontational (or, depending on one's point of view, more realistic) stance toward the USSR. Moreover, it is easy to see how these principles and goals fit directly into Reagan's personal beliefs concerning foreign and defense policies.

Space Commerce

Goals three and four of NSDD 42 were "obtain[ing] economic and scientific benefits through the exploitation of space" and "expand[ing] United States private-sector investment and involvement in civil space and space-related activities." For the first time in the history of the U.S. space program, a high-level official document made a direct reference to the American business community. Of course, the idea of space's "economic benefits"—communications, weather forecasting, remote sensing, navigation, and so forth—was nothing new. As previous chapters have noted, these had been part of the discussion surrounding U.S. space policy (albeit usually not a very important part) as far back as the 1950s. Most space advocates, however, seldom addressed specifically the question of exactly how these services would be provided, or by whom. Up until this time, space activity had been undertaken almost exclusively by governments,[103] and at least until the Reagan presidency, there was little indication that this was going to change. With NSDD 42, the Reagan White House began the process of challenging, and ultimately overcoming, the assumption that all space service delivery had to be through government provision. Indeed, by 1988, Secretary of Commerce William Verity would be referring to space as "just a place to do business."[104]

As was the case with national defense, the commercialization issue was well represented throughout NSDD 42. Another "basic principle," for example, was that "[t]he United States encourages domestic commercial exploitation of space capabilities, technology, and systems for national economic benefit."[105] Further down, under a section entitled "Civil Space Program," the directive declared:

> The United States government will provide a climate conducive to expanded private sector investment and involvement in civil space activities, with due regard to public safety and national security. Private sector space activities will be authorized and supervised or regulated by the government to the extent required by treaty or international law.[106]

Finally, the NSDD called for making the space shuttle available to all commercial users (with the usual national security proviso), and for the eventual privatization of government remote-sensing satellites.[107]

A follow-up to the directive (National Security Study Directive 13-82, issued the following December) continued this theme. It called for an inquiry into the organization, roles, and responsibilities of "[t]he Private Sector [sic], especially concerning its relationship with the U.S. Government and the need for regulation, oversight, and incentives to stimulate investment," and for the government to identify "new areas of private sector investment in space which the Administration should stimulate."[108] An Issue Paper on the United States Space Program, written by the Office of Policy Development in February 1983, concluded that "[t]he future will bring further exploitation of unique advantages afforded by space, with particular emphasis on . . . private sector participation."[109] In general, like the approach to military applications, these statements on space commercialization were completely consistent with the administration's overall philosophies concerning private enterprise and the proper role of government. And as his subsequent public comments on the matter would make abundantly clear, they were also in perfect accord with President Reagan's personal views.

Redefinition and Political Rhetoric

Although the White House clearly considered both space defense and commercialization to be high priorities, for obvious reasons there was far more emphasis placed on the latter in official statements, speeches, and public events. In August 1983, for example, the White House invited representatives from 11 major space-related companies to meet with high-level administration officials and have lunch with the president. The primary purpose of the conference, which was widely reported in the trade press,[110] was to explore the "major themes" related to encouraging more private enterprise activity in space. Topics included economic incentives, techniques for expanding space markets, removal of regulatory barriers, and the appropriate role for NASA.[111] There was also some discussion of whether the government should build a space station, with some of the industry participants reportedly making an appeal for such a facility to Reagan directly.[112] Ongoing discussion with business leaders was to play a major role in shaping the administration's first major policy changes regarding commercial space.

For his part, the president seldom missed an opportunity to hail the economic potential of space. Just as Kennedy, Johnson, and sometimes even Eisenhower would speak of the U.S. space program largely (if not exclusively) in the context of an American–Soviet "race" during the late 1950s and early 1960s, Reagan almost never made any statement about space policy without also mentioning—usually with great enthusiasm—the whole host of new products, additional marketing opportunities, and potential for job creation that it would bring about.

During his 1984 State of the Union Message, for example, the president declared:

130   DEFINING NASA

> Just as the Oceans opened up a new world for Clipper Ships and Yankee Traders, space holds enormous potential for commerce today. The market for space transportation could surpass our ability to develop it.... We'll soon implement a number of executive initiatives, develop proposals to ease regulatory constraints, and, with NASA's help, promote private sector investment in space.[113]

The president chose the 15th anniversary of the first moon landing to announce his new commercial policies. Proclaiming that *Apollo XI* "wasn't our last great moment in space," but that "most of our great moments are ahead of us," Reagan called for encouraging private investment in space in order "to improve the quality of life on earth." Given the opportunity, space-based businesses could perhaps find cures for diabetes—or even cancer—"create new metals that are lighter and stronger than anything we've ever known," generate "tens of thousands of jobs, billions of dollars in foreign trade, and tens of billions of dollars added to the gross national product."[114] The following year, at a speech at the National Space Club luncheon, the president continued with this theme:

> Individual freedom and the profit motive were the engines of progress which transformed an American wilderness into an economic dynamo that provided the American people with a standard of living that is still the envy of the world.... We must make sure that the same incentives that worked so well in developing America's first frontier are brought into play in taming the frontier of space.[115]

Of course, talk—even such highly extravagant talk—is cheap. As the introduction to chapter 5 pointed out, political statements of intent are truly meaningful only to the extent that they are supported by specific policies and programs. As it happened, by the time Reagan delivered his address to the National Space Club, a number of far-reaching initiatives were well underway.

Step Two: A New Policy Agenda

Consistent with their new definitions of space policy, Reagan officials soon set out to reshape the program to an extent not seen since the early days of Project Apollo. Some of the proposals and initiatives were blocked or canceled by Congress, others were opposed even by some members of the administration, and not all of them could be counted as entirely successful (a common occurrence for government programs involving emerging technologies). Nevertheless, they stand as compelling evidence that the White House was fully committed to their particular vision of what a public space program was all about.

Defense

As already noted, the frequency of public statements concerning the new approach to military space came nowhere near matching that associated with space commerce. This certainly did not mean, however, that the area was being neglected. In fact, the administration oversaw a major expansion of the defense space sector in almost all areas. Two major programs, both of which involve development of new space technologies, merit particular attention.

Strategic Defense Initiative. The most well-known—and controversial—project associated with Reagan-era space-based defense policy was the Strategic Defense Initiative (SDI), a plan to construct a "shield" to protect the United States from a missile attack. Most Americans heard of the proposal for the first time during a televised speech by the president on March 23, 1983, in which he called on "the scientific community . . . to turn their great talents now to the cause of mankind and world peace, to give us the means of rendering [nuclear] weapons impotent and obsolete."[116]

The idea, however, was by no means new. It had been widely discussed in conservative political circles during the 1970s,[117] and Reagan had personally become acquainted with the concept of ballistic missile defense (BMD) during a tour of the Lawrence Livermore National Laboratory in 1967, shortly after he had been elected governor of California.[118] He reportedly had wanted to make BMD a major issue in the 1980 presidential campaign, but was talked out of it by aides who feared its association with the Republican right wing[119] (it was, however, included in the 1980 GOP platform[120]).

Indeed, this appears to have been one of those issues in which the president parted company with many of his closest aides. Although the specific proposal had been developed by Robert McFarlane and Admiral James Watkins, who was then chief of naval operations, it was one of the few initiatives in his presidency for which he personally—and uncharacteristically—took sole credit: "SDI was my idea," he told reporter Lou Cannon in 1989,[121] a claim that has been borne out by the president's aides as well as outside observers.[122] Very few administration officials were even told in advance that the proposal was forthcoming (the ostensible purpose of the March 23 address was to talk about defense spending),[123] and support for the idea among most of them was tepid at best, although all were duty bound to support the president, at least in public.[124] Nevertheless, some White House officials opposed to SDI (reportedly including Chief of Staff James Baker) undertook behind-the-scenes efforts to "moderate" the proposal.[125]

That the measure not only survived, but flourished—appropriations for SDI had totaled more than $15 billion by the end of the president's second term and $36 billion nearly a decade later (under a Democratic president, no less)[126]—was due in no small part to Reagan's direct interest in it.[127] Members of Congress and officials in the federal bureaucracy, even if they were themselves skeptical of the idea, were all aware that the president was personally following the progress of the issue. This (relatively) close attention helped ensure that SDI outlasted most of its opponents. What is even more impressive is that the program lived beyond not only the Reagan Presidency, but the end of the Cold War itself, and continues on even to this day.

Anti-Satellite Systems. The other major—and also highly controversial—program of this period was the renewed emphasis on antisatellite systems (ASAT). As satellite technology became an increasingly important element in military planning and operations (for reconnaissance, communications, and navigation), defense officials began to fear that satellites themselves could become a military target. The USSR was known to have begun experimenting with ASAT technology in the late 1960s, and had conducted around 20 tests between 1968 and 1982.[128] The Soviets were apparently attempting to develop what was known as a "co-orbital" ASAT, in which an explosive device (a so-called killer satellite) is launched into an orbit designed to intersect with its target. Once within range, it is detonated.

Although there was a U.S. ASAT program in place by the time Reagan took office, previous administrations had not given it very high priority. President Carter had even gone so far as to open up the possibility of negotiations with the USSR on a treaty to limit, or even to eliminate, such systems.[129] President Reagan, however, was anxious to move forward with ASAT development. NSDD 42 declared:

> The United States will proceed with development of an ASAT capability, with operational deployment as a goal. The primary purposes of a United States ASAT capability are to deter threats to the space systems of the United States and its Allies and, within such limits imposed by international law, to deny any adversary the use of space-based systems that provide support to hostile military forces.[130]

The Reagan administration was therefore the first to consider ASAT capability as a form of deterrence.[131] Moreover, it seems clear from the latter part of the section that the administration was not adverse to using such a system in a "first-strike" so as to "deny any adversary the use of space-based systems."

The U.S. Air Force conducted two successful antisatellite tests in 1984. Unlike the Soviet co-orbital approach, the American weapon was carried on a small rocket, fired from an F-15. After these two successes, however, Congress banned further ASAT research, largely because the Soviet Union had been adhering to a unilateral moratorium on ASAT development since 1982.[132]

⊙ ⊙ ⊙

In general, funding for defense-related space research and development doubled during Reagan's first term, while that for ground support (satellite detection, tracking, and control) tripled.[133] Overall, DOD spending on space increased from $3.8 billion to $17.7 billion—growing from 44 percent to 66 percent of the total federal space budget—between 1980 and 1988. By contrast, over the same period, NASA's share of government space spending fell from 54 percent to 31 percent. In other words, by the end of Reagan's second term, the nation's leading space agency (at least in budgetary terms) was DOD, not NASA.[134]

Space Business

Beginning in 1984, the White House proposed, instituted, and administered a large number of new programs intended to carry out the edicts of NSDD

42. It is particularly interesting to note how closely many of these initiatives match up with the so-called grand themes of the Reagan presidency discussed earlier.

Taxation/Regulation. In his radio address commemorating the *Apollo 11* anniversary, Reagan spoke of the need to remove "needless regulation" and to revise existing tax and tariff laws that "inadvertently discriminate against companies that do business in space rather than on the ground":

> For example, the way the law is written now, products made in space might be subject to import tariffs just because they weren't made in America. Well, we're going to change that. Another example: Businesses which operate at home receive various kinds of tax incentives. But, again, as the laws are written now, space products companies would not receive those incentives. We'll be looking at that, too.[135]

Accordingly, the administration issued NSDD 144, entitled "National Space Strategy," that called for the federal government to "encourage the private sector to undertake commercial space ventures" by eliminating or revising "discriminatory" laws and regulations, and by updating laws and regulations "to accommodate space commercialization."[136]

The most clear-cut—and easily the most successful—application of this principle was with regard to the commercial launch industry. By the 1980s, a number of private firms (particularly in the United States) had developed considerable experience in building and servicing rocket boosters for government and military organizations. Nevertheless, space launches had remained exclusively a government function. In 1982, however, the first privately developed commercial rocket, *Conestoga I*, was successfully test launched by Space Services Incorporated, a firm based in Houston. SSI's president, David Hannah, informed Congress the following year that obtaining the necessary licenses and waivers was the company's biggest single expense in the entire project.[137] Essentially, the firm had to satisfy the demands of 22 different federal statutes, as well as deal with 18 separate agencies, including NASA, the Federal Aviation Administration (FAA), the Federal Communications Commission (FCC), the State Department, and the Bureau of Alcohol, Tobacco, and Firearms (ATF).[138] Most (if not all) of these requirements had been formulated in a piecemeal fashion by each individual organization, without any regard for how they might possibly affect the commercial launch sector (which, after all, did not even exist when the regulations were created).

The Reagan administration's efforts to ease this regulatory tangle—and to help in the development of a new industry—started even before NSDD 144 was issued. The previous February, the president signed an executive order in designating the Department of Transportation (DOT) as the "single point of contact" to "expedite the processing of private sector requests to obtain licenses" to operate expendable launch vehicles.[139] In other words, instead of applying for separate permits and waivers from the FCC, NASA, the ATF, and the others, commercial launch vendors would simply go to a single agency housed within the Transportation Department.

That same year, Congress passed the Commercial Space Launch Act of 1984, which codified most of the provisions in the executive order.[140] In 1988, Congress amended the act to allow the secretary of transportation to set the maximum level of insurance required for private launch providers using government-owned launch facilities.[141] The first two launches licensed by DOT (conducted by SSI and McDonnell Douglas) took place in 1989,[142] and by 1992, the U.S. commercial space launches were generating more than $500 million in revenue.[143] In fact, by the end of the 1990s, the industry had matured to the point where most government payloads—including those for the military—were being placed in orbit by private launchers.[144]

Privatization. Obviously, the creation of the commercial launch industry does represent a type of privatization. In addition, however, the administration did attempt—this time with rather less success—to engage in the type of "load shedding" described above. That is, it tried to transfer selected government-owned, space-related assets to the private sector. These efforts by and large failed for both political and economic reasons.

In 1981, for example, Reagan sought to privatize the satellite system—managed by the National Oceanic and Atmospheric Administration—that provides data for the National Weather Service. A majority of the Congress, however, felt strongly that information about the weather represented a public good, and should therefore be obtained and disseminated by the government. It moved, therefore, to block the satellite sale, going so far as to include a specific prohibition on any such transfer as a provision in the FY 1984 budget.[145]

A second attempt at satellite privatization, however, proved to be more successful, at least initially. For many years, the technology known as remote sensing had been steadily, if slowly, moving into the civilian marketplace.[146] Basically a derivative of military reconnaissance satellites, remote sensing involves the use of high-resolution photographs of the earth from space to aid activities such as map-making, prospecting for oil or other minerals, and making crop forecasts. During the 1970s, the federal government launched a number of satellites, known as "Landsat," for these and other purposes.

At the administration's urging, Congress passed the Land Remote-Sensing Act of 1984, which authorized the secretary of commerce to award to private industry a contract to market Landsat data.[147] The contract was awarded to EOSAT, a consortium of Hughes Aircraft Corporation and RCA. Unfortunately, for a variety of reasons—both political and market related—this arrangement was essentially rescinded eight years later, when the federal government took over responsibility for the development of Landsat 7.[148]

Finally, in addition to the nation's satellite systems, the administration also sought to move part of its own space station program to a private space facility. This program (and its problems) will be discussed in the next chapter.

Providing Infrastructure: Aid to "Infant Industries." Although, as noted above, economic conservatives generally want government to be removed from most economic decisions, most acknowledge some role for the state in creating and maintaining national infrastructure, that is, the network of roads, tunnels, bridges, communication networks, and so on that allow goods and services to move freely, but that are not the province of any one business or industry. In general, it is held, government intervention is required to assure free access to these facilities and to ensure that they are kept in reasonably good condition.

Such thinking can be seen throughout space policy in the 1980s. In the words of one analyst:

> If we were in Canada and one were to turn to industry and say: I think it is a good idea if you build a widget plant and the industrialists say yes; widgets are selling and we should build a widget plant. And then we say to them: we want you to build it in the middle of the Northwest Territory. The businessman asks why? There are no roads, there is no electricity, there is no gas, no water, no sewage, nothing. That is the role of government, municipal or federal. The role of government is to provide the infrastructure necessary for business to operate. The same thing is true in space. We are asking business sometime in the future to build machinery to do business in space. It is the responsibility of government to put that infrastructure in space.[149]

Thus, according to early administration policy (e.g., NSDD 42), the shuttle was to have been the primary means—virtually like a highway—for business and government to get into space, and NASA was directed to ensure that all legitimate users were to have guaranteed access to it. In addition, viewing the shuttle in this way justifies the public subsidies associated with its (pre-1986) use. Similar arguments were used by proponents during the internal administration debate over approving NASA's request for a space station (see below).

Along these same lines, the administration was sometimes willing to provide some forms of aid to firms that were engaged in developing newer, "cutting-edge" technologies (sometimes known as "infant industries"). As noted earlier, policies to assist "technologies requiring a longer period of initial development" had been in place as early as 1982. For the most part, the administration sought to support these new ventures without resorting to direct subsidies. Indeed, the notion of subsidizing private space enterprises was explicitly rejected in most commercialization documents. There was, however, an important exception to this proscription: until the *Challenger* accident in 1986, commercial users of the space shuttle were charged rates that were far below its "true" costs (or, in the language of the time, shuttle pricing did not reflect "full-cost recovery"). Again, more will be said about this issue below and in the following chapter.

Generally, however, the government limited its help to fledgling space industry to simply becoming its primary (if not its only) customer. This approach served not only to provide the company with revenue, but also to send an important signal to other potential customers or investors. Once again, the commercial launch sector is the most prominent example. Even as late as 1992, when the industry's revenues were in the neighborhood of $500 million, government users accounted for nearly 90 percent of its sales.[150]

Step Four: Reorganization

In view of all of this activity, it should come as no surprise that the Reagan administration was responsible for the most extensive overhaul of the organizational structure of the American space program since the *Sputnik* era. As might also be expected, this reorganization was largely (although not exclusively) policy driven (as opposed to some of the other forces that can motivate reorganization as described in chapter 2). What is somewhat surprising—albeit virtually unnoticed at the time—was where this extensive reshuffling ultimately would leave NASA.

To help set the broader parameters for space policy, the administration made extensive use of a "cross-disciplinary" approach to executive organization that had also been employed when Reagan was governor of California.[151] Recognizing that the responsibility for space-based activity had spread well beyond NASA and DOD (by the early 1980s, it also included the Departments of Commerce and State, the CIA, and several others), NSDD 42 established a Senior Interagency Group for Space, usually referred to as SIG (Space). Its members included the administrator of NASA; the deputy or undersecretaries of state, commerce, and defense; the director of the CIA and the Arms Control and Development Agency; and the chairman of the Joint Chiefs of Staff. It was chaired by the national security advisor, with the science advisor and budget director serving as nonvoting "observers."[152]

In a somewhat similar vein, the traditional cabinet structure was complemented by a unique system of "cabinet councils," five cross-departmental boards responsible for major areas of public policy: Economic Affairs, Commerce and Trade, Human Resources, Natural Resources and the Environment, and Agriculture and Food. The Cabinet Council on Commerce and Trade (CCCT) quickly became the administration's focal point regarding issues related to space commercialization and privatization. In addition, the Economic Policy Council would at one point become involved with a specific space commerce issue concerning NASA's space station (see discussion below and in the next chapter).

Even before NSDD 42 was issued, Craig Fuller, assistant to the president for cabinet affairs and a staff member of CCCT, had been working to incorporate such initiatives into the proposed National Space Policy. In a June 1982 memo (i.e., the month before the new policy became public), for example, Fuller raised a series of questions about space and the private sector, including whether the National Aeronautics and Space Act should be amended to allow NASA a greater role in fostering space commercialization, the "level of . . . private resources" needed to maintain a U.S. lead in the space area, and "what sort of tax incentives should be considered with respect to encouraging greater commercialization."[153] CCCT also played a role in setting up the administration's commercial space launch policy.[154] Finally, in April 1984, CCCT established a Working Group on Space Commercialization, the final report of which (issued in June) provided much of the material for Reagan's July speech on commercial space initiatives.[155]

In addition to these less formal executive branch organizations, which generally change from one presidential administration to the next,[156] the Reagan administration and Congress also established a number of new permanent organizations designed to implement specific programs. To carry out its new responsibilities for regulating the private launch industry, for example, DOT in 1984 created an Office of Commercial Space Transportation. In 1987, the Department of Commerce, which had been involved in the administration's commercialization efforts from the beginning (Reagan's first commerce secretary, Malcom Baldridge, was chairman of CCCT),[157] set up an Office of Commercial Space to encourage the private development of space technologies and applications.[158] Within DOD, the U.S. Air Force Space Command was established in December 1982 to oversee the operation of the (rapidly expanding) U.S. space defense system.[159] Finally, SDI research and development was administered by a new Strategic Defense Initiative Office.

And NASA?

It may well appear that the discussion up to this point has departed significantly from the main theme of this book, which is the evolution of NASA's mission. It is therefore worth asking precisely how—or even if—Reagan officials saw the nation's first and still most important space agency fitting into their newly defined space policy. Obviously, under the terms of the National Aeronautics and Space Act of 1958, NASA could have no direct role in the administration's expanded programs of space-based defense: by law, that is the exclusive province of DOD. With regard to space commerce, however, there clearly were opportunities for NASA to participate (the place of scientific research, another of the agency's primary responsibilities, in all of this will be considered in the next chapter). It had, after all, been directly responsible for the development and dissemination of virtually all of the technologies that were now being discussed as part of the commercialization effort (rocket launchers, space platforms, various types of satellites, etc.). Moreover, ever since its major post-Apollo initiatives had been rejected (see last chapter), NASA on its own had been looking for ways to move more directly into providing commercial services—that had, in fact, been one of the primary justifications for the space shuttle program.

To be sure, there appeared to be many in the administration and Congress who wished to see NASA move in just such a direction. Chief among these was Reagan himself. On a number of occasions, the president gave the clear impression that he was quite favorably disposed toward the agency. In a speech commemorating its 25th anniversary, for example, he declared that "NASA's done so much to galvanize our spirit as a people, to reassure us of our greatness and our potential."[160] As has already been seen, he also made frequent references to its great achievements of the past, particularly the Apollo moon landings. More to the point, Reagan often noted that his new initiatives on space transportation or privatization would be carried out "with NASA's help."[161] Finally, he gave his support to NASA's largest program of the decade, the space station (see below).

Most of the early administration space policy documents—both public and internal—also acknowledge the agency as the principal organization in U.S. space policy, suggesting that officials expected, at least initially, that NASA was to play a significant role in implementing their programs. NSDD 42, for example, gives the agency responsibility for "operational control of STS for civil missions" and for assuring "the Shuttle's utility to the civil users [sic]."[162] Along these same lines, NASA was heavily involved in the CCCT Working Group on Space Commercialization, serving as the "convening" agency at each of its meetings and in the drafting of its final report.[163] Not surprisingly, that document repeatedly "reaffirms" that NASA is the "lead agency for space non-regulatory functions" and calls for it to continue working toward providing low-cost access to near-earth orbit and otherwise help to produce an environment conducive to private sector development.[164]

Finally, Congress took its own steps to make commercialization a permanent part of NASA's operation. In 1984, it passed an amendment to the National Aeronautics and Space Act declaring that "the general welfare of the United States requires that the National Aeronautics and Space Administration seek and encourage to the maximum extent possible the fullest commercial use of space."[166]

Putting NASA in Business

Acting in accordance with these directives and recommendations, NASA set out that same year to develop its own Commercial Space Policy, based on five operating principles:

1. The Government should reach out to and establish new links with the private sector.

2. Regardless of the Government's view of a project's feasibility it should not impede private efforts to undertake commercial space ventures.

3. If the private sector can operate a space venture more efficiently than government, then such commercialization should be encouraged.

4. The Government should not expend tax dollars for endeavors the private sector is willing to underwrite. However, the Government should invest in high-cost and/or high-risk technologies and space facilities which encourage private investment.

5. When a significant Government contribution to a commercial space endeavor is requested, generally two requirements should be met. First, the private sector must have significant capital at risk, and second, there must be significant potential benefits for the nation.[166]

Over the next few years, NASA set out to implement its new policy through a variety of channels. It continued to offer commercial users relatively inexpensive (i.e., below cost) access to the shuttle. With the international market for space launches becoming increasingly dominated by the European Arianne rocket, this practice had the added beauty of giving the agency a role in the growing political debate over American economic competitiveness (which, as noted earlier, was also a concern of the President's[167]).

NASA also began trying to stimulate the development of new commercial space technologies. In 1985, it established a number of Centers for the Commercial Development of Space (CCDS). These government–industry–university partnerships were to use "seed" money from the agency (ranging from $750,000 to $1.1 million[168]) to promote the creation of new industries in such areas as materials processing, life sciences, remote sensing, automation and robotics, propulsion, structures and materials, and power sources. NASA expected that the centers would evolve into self-sufficient business enterprises within an average of five years.[169] Starting with five in 1985, the number of CCDSs had increased to 17 by 1990.

Along these same lines, the agency made efforts to "reach out and establish new links with private business." In 1988, for example, NASA signed an agreement with Spacehab, Inc., a Washington-based company promoting industrial research in microgravity processing. Under the agreement, the company was to develop a pressurized module, to be flown inside the shuttle's main cargo bay, containing 50 "lockers" for supporting a variety of microgravity experiments.[170] These lockers—which would be tended by shuttle astronauts who would move to and from the module through a tunnel connected to the shuttles mid-deck—would be leased to government and commercial users. For its first six flights (which began in 1992) NASA committed itself to leasing 200 of the 300 lockers available.[171]

The New Transcontinental Railway?

Perhaps the most telling example of NASA's new "commercial approach" was the manner in which it and its allies in the administration secured presidential approval for the agency's most sought-after project, a permanently inhabited space station. Ever since Apollo, NASA's efforts to begin work on such a facility had been repeatedly rejected as simply another high-cost space "stunt" that the nation could not afford.[172] Until 1983, policymakers simply could not be persuaded that an orbital station served any pressing national need. This view continued into the early years of the Reagan administration. James Beggs and Hans Mark, the two top men at NASA, were both strong supporters of a station, and Beggs had even discussed the idea during his Senate confirmation hearing.[173] To present the idea to the president, however, it first had to receive the endorsement of SIG (Space). Unfortunately for NASA, the interagency group, which was heavily skewed in the direction of national defense (five out of its eight voting members came from security-related organizations), refused to pass on the project, primarily due to opposition from the Defense Department. Although (as seen in previous chapters) DOD had sought repeatedly to get military personnel into space in the early days of the program—and had even pursued its own space station program, the Manned Orbiting Laboratory, in the 1960s—by the 1980s the department had concluded that its mission could be conducted more efficiently (not to mention more cheaply) with unmanned satellites.

In retrospect, it is a matter of supreme irony that a proposal long derided by its opponents as one more example of a big-spending public program would not only receive the blessing of a president who was generally committed to shrinking the size of the federal government, but would also obtain such approval largely on the basis of its commercial appeal. As noted above, Reagan was first introduced to the idea at the August 1983 space commercialization conference, where a number of business leaders told him that they needed an orbital facility to develop new products and create new industries, but could not afford to build one on their own.[174] That same month, the president received a letter from congressional supporters claiming that a station was "particularly compatible with your economic program" and that "without government backing in this largely uncharted area, space development will be unnecessarily delayed."[175]

By far, however, NASA's most important partner in this endeavor was CCCT. In crafting its commercial space proposals (with, as noted earlier, significant NASA input), the council had declared that a permanently inhabited orbital facility was squarely within the government's "infrastructure" responsibilities. Moreover, CCCT believed that one of the more serious impediments to commercial space development was the perception within the business community that government policy in this area was constantly shifting. According to this view, private enterprise was reluctant to make substantial investments—and to shoulder the attendant financial risks—in space enterprises without some assurance that federal support would continue for more than a few years (i.e., beyond the span of a single presidential administration). Thus, in addition to its direct utility, a space station would provide potential investors and entrepreneurs with a clear and dramatic demonstration of the government's commitment to their operations.[176]

Besides providing them with much-needed political ammunition, CCCT was able to give station advocates something that, from a practical point of view, was all that really mattered: access to the president. The most obvious route to this goal, SIG (Space), had been closed when that body refused to give their endorsement. It was therefore at a meeting of the Cabinet Council on Commerce and Trade, on December 1, 1983, that proponents finally had the opportunity—complete with a scale model and view graphs—to make a direct appeal to Reagan personally.

Using CCCT as a forum had the added advantage of broadening the discussion beyond the national security interests so heavily represented in SIG (Space). Of course, a DOD representative was present—as were such station opponents as Stockman and Keyworth—but the meeting also included supporters like trade representative William Brock, Commerce Secretary Malcolm Baldridge, Craig Fuller, James Beggs, and Hans Mark. It was, as Howard McCurdy has described it, "the first time in the interagency review process where . . . the ayes at least equaled the nays."[177] It was also of no small importance that the makeup of this group served to highlight the facility's commercial potential. Although Reagan did not announce his decision at the meeting, NASA clearly had carried the day.

Of course, NASA viewed the station as far more than simply a place to conduct business. It was also supposed to serve as a laboratory for basic research in biology, space science, geophysics, and many other disciplines.[178] For many in the agency (and spaceflight advocates generally), however, its most important feature was that it seemed to signal a renewal of the human exploration of space. With the station as a "staging area," it was possible once again to envision such grand projects as a permanent base on the moon and, eventually, missions to Mars. If fulfilling that long-cherished dream required a certain amount of profit-making activity along the way, so be it.

In other words, just as supporters of an aggressive government space program had been perfectly willing to couple their ambitions with the Cold War fears of policymakers and the general public during the 1960s,[179] NASA was seeking to graft its larger goals onto the economic-based priorities of the 1980s. Stated simply, the agency was making every effort to carry on as before, albeit under a new definition of space policy.

For a brief time, it appeared as though NASA's hoped-for marriage of exploration and profitability might actually succeed. While clearly committed to his commercialization policies, President Reagan had always been a firm believer—at least judging from his rhetoric—in the space program's loftier goals.[180] Rather than simply issuing a statement supporting the station, the president opted to unveil the project in the most public fashion available, his annual State of the Union Message. On January 25, 1984, in the same spot where President Kennedy had issued his lunar landing challenge, Reagan announced the new initiative in terms that deftly combined the lofty with the practical:

> The Space Age is barely a quarter of a century old. But already we've pushed civilization forward with our advances in science and technology. Opportunities and jobs will multiply as we cross new thresholds of knowledge and reach deeper into the unknown. . . .
>
> America has always been greatest when we dared to be great. We can reach for our greatness again. We can follow our dreams to distant stars, living and working in space for peaceful, economic, scientific gain. Tonight, I am directing NASA to develop a permanently manned space station and to do it within a decade.
>
> A space station will permit quantum leaps in our research in science, communications, in metals, and in lifesaving medicines which could be manufactured only in space.[181]

Needless to say, NASA officials and their supporters were ecstatic.[182] Their mood, however, would prove to be short-lived. Despite the president's endorsement (and continued support), the agency's bid to redefine its mission, and to link its other aims to it, was about to receive several serious setbacks. It remains an open question as to whether it has, or ever will, fully recover.

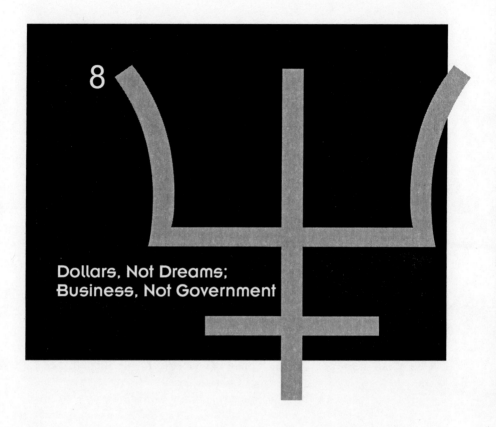

8
Dollars, Not Dreams; Business, Not Government

For over 40 years, succeeding generations of space enthusiasts had been able to secure the resources necessary to advance the state of the art in spaceflight technology—at times considerably—by convincing a wide variety of political leaders that it would help them achieve their own goals. Wernher von Braun and his colleagues were able to develop the first liquid-fueled rocket only because Adolf Hitler was persuaded that it could be an effective weapon of war. Nearly 20 years later, believing that only by beating the Soviets to the moon could the United States demonstrate its true resolution and spirit, American policymakers lavished unprecedented amounts of money, attention, and prestige on NASA scientists and engineers. A decade after that, NASA officials received presidential and congressional approval—and, once again, billions of dollars—to build the world's first reusable spacecraft by convincing Richard Nixon and the Congress that such a vehicle would allow economical access to earth orbit (and that it would help win the Republican Party votes in aerospace-dependent states like California).[1]

By the beginning of 1984, it was starting to look as though they had done it yet again. The most conservative president in modern American history, one passionately devoted to shrinking the size of government and reducing public spending, had just been persuaded to support one of the largest space initiatives—indeed, one of the largest public works projects—ever, largely on the strength of the claim that it would provide a boost to private enterprise. NASA was actively seeking business customers for its growing fleet of shuttles, and was busily promoting new space-based enterprises all across the country. The agency, it seemed, had successfully beaten its Cold War swords into plowshares, and was poised to use this position to continue its own goal, advancing humanity into space.

This time, however, the transformation could not be sustained. Just four years after President Reagan's dramatic space station declaration, NASA's status, and its overall role in U.S. space policy, would, if anything, be *less* clear than it had been before Reagan took office. What makes this situation particularly ironic is the fact that by the end of the president's second term, the actual use of space technology was growing faster than ever before.

What could account for such a complete (and rapid) turnaround? Not surprisingly, many commentators blame NASA itself, claiming that by this time the agency had become too political and/or too bureaucratic or too fixated on "big" missions like the space station.[2] A somewhat more comprehensive view of this period (which does not necessarily minimize NASA's responsibility), however, requires placing the agency's actions in the context of the changing definitions of space policy. When viewed in this light, the decline of the agency in the late 1980s can be attributed to a number of factors, both external and internal.

The New Politics of Space Commerce

In retrospect, it now seems perfectly obvious: public officials who decry "big government" programs and call for bureaucracies to "get out of the way" of the private sector usually do not look favorably on an agency the size and scope of NASA, even when it seems to be (some might even say *particularly* when it seems to be) supporting their agenda. Seen in this context, the intent of official documents like NSDD 42 was not—as many saw it at the time—to develop a new role for NASA, but rather to remove it (and all other government organizations) from space activity to the greatest extent possible.[3]

Traditional space advocates could perhaps be forgiven for not immediately grasping the full meaning of this new political reality. After all, in the early 1980s this represented a radically different way of thinking about space: as the last chapter noted, up to this time virtually all space-based services had been provided by government. Moreover (as also seen in the last chapter), space technology possesses a number of features—the development of large-scale infrastructure; cutting-edge, high-risk research; and so on—that even many economic conservatives argue requires some role for government.

Nevertheless, it is clear that by the middle of the decade, many people in the administration (as well as in Congress and in the private sector) began to view NASA as an impediment to their plans. The resulting controversy quickly grew to engulf two of the agency's largest, most expensive, and most visible programs.

STS Pricing

By far, the most serious of these conflicts centered on NASA's policies toward commercial users of the space shuttle. As noted earlier, one of the agency's primary justifications of STS was that it would provide private business with "routine and economical" access to space. Several critics, however, began pointing out that shuttle flights were only "economical" because NASA underpriced its services while making up the difference out of the STS budget.[4] In other words, commercial shuttle flights were based on hefty government subsidies.

According to one estimate, were NASA to charge nongovernment users a price based on so-called full-cost recovery, that is, a price that reflected the actual expenses involved in operating the shuttle, businesses would end up paying somewhere in the neighborhood of $85 million per flight. The agency, however, was only charging its commercial customers around $20 million. By contrast, private launch companies, who were using less expensive expendable launch vehicles, needed to charge $25–30 million in order to turn a reasonable profit. This figure was far less than the shuttle's real cost, but millions of dollars higher than the subsidized price that NASA charged commercial users.[5] Under such conditions, private firms would have a very difficult time (to say the least) competing with NASA.

This practice had come under some criticism within the Reagan administration as early as 1982.[6] In the first few months of 1983, after only six shuttle flights (including the first by *Challenger*, the second orbiter to join the fleet), free market advocates were complaining that the policy was detrimental to the development of the private sector launch industry. A report from the conservative Heritage Foundation, for example, warned that the federal government would "suffocate space entrepreneurs with red tape and subsidized government competition."[7]

By mid-1984, as the administration's commercial launch policy was moving forward, NASA's pricing policies came under renewed scrutiny. In late June, Representative Ed Zschau of California urged his Republican colleagues in Congress to sign a petition for President Reagan protesting NASA's practice of using "tax dollars to market its launch services." This policy, according to the congressman, not only threatened to "devastate a budding private sector industry," but was also "clearly inconsistent with the principles of our party [and] the statements made by the president in his State of the Union address"[8]:

> We continue to support whole-heartedly NASA in its quest for new discoveries. However, in its eagerness to provide greater justification for the space shuttle, we fear that the agency may be diluting its basic R&D mission to include providing commercial services in competition with the private sector. NASA should recognize that the real incentives for commercializing space lie in a private sector environment that encourages entrepreneurship, and that such an environment can be greatly damaged by unfair competition on the part of a federal agency.[9]

One of the more significant features of this petition is its assessment of NASA's motives. Phrases like "in its eagerness to provide greater justification for the space shuttle" suggest that at least some in Congress believed (not without reason) that the agency was using the politically popular notion of expanding space commerce as a means of advancing its own agenda. Such a charge would be heard more than once in the years ahead.

Meanwhile, the issue was also receiving a complete airing within various executive branch agencies. Not surprisingly, the Department of Transportation (which had been assigned responsibility for the commercial launch industry) was strongly opposed to the NASA policy, and supported the removal of government subsidies. This was also the position taken by the Office of Management and Budget, which noted that its own guidelines prohibited government from providing a commercial product or service that could be produced more economically by the private sector.[10]

NASA, of course, argued on behalf of the current shuttle pricing system. In a series of letters to the White House, Administrator James Beggs asserted that "pricing the shuttle out of the market" would only cripple the program without necessarily aiding industry. Rather, stated Beggs, the most likely beneficiary would be the European Arianne launcher (which itself received significant external subsidies). Thus, an increase in shuttle prices could have a serious effect on U.S. international economic competitiveness. Moreover, Beggs maintained, the relatively low shuttle prices encouraged a number of businesses to enter into space-based initiatives in the first place.[11]

Given its stated support (not to mention all of its legislative and administrative activity) for the commercial launch industry, it seems somewhat surprising that the White House chose to side with NASA. In a memo to DOT Secretary Elizabeth Dole, Robert McFarlane, the president's assistant for national security affairs, stated that "we must proceed prudently and cautiously in resolving this issue."[12] It cited the problem of international competitiveness, as well as the need to preserve the "unique attributes" of the shuttle.

Thus, in August 1984, the administration issued a new National Space Strategy allowing NASA to postpone implementing a full-cost recovery pricing structure for the shuttle until October 1, 1988 (i.e., the beginning of FY 1989).[13] In another National Security Decision Directive issued the following July, the post-1988 minimum price was set at $74 million (1982 dollars),[14] a figure that many in DOT felt was still too low (relative to the shuttle's actual cost).[15] Nevertheless, the matter was apparently settled. It would, however, be reopened in an unexpected and tragic fashion six months later with the launch of the space shuttle *Challenger* (see below).

A Private Space Station?

Although easily the most visible, shuttle pricing was by no means the only instance where policymakers began to regard a major NASA program as a barrier to free enterprise. Less than two years after the STS controversy, a similar type of dispute erupted between the agency, the rest of the administration, and some influential members of Congress, this time over what NASA viewed as its most important project, the space station.

In 1984, Space Industries Incorporated (SII), a Houston-based firm (headed by Maxime Faget, a retired NASA employee who had developed the original concept for Project Mercury), approached the government with a plan to build the Industrial Space Facility (ISF), styled as an "industrial park in space."[16] SII planned to lease space on the ISF to private companies and the federal government to conduct long-range experiments in such areas as low-gravity materials processing.[17] Unlike NASA's future space station, which was to be continuously occupied, the ISF was only to be "human tended," meaning that shuttle astronauts would visit it two or three times a year to replace experiments and conduct routine maintenance. Day-to-day operation of the facility would be managed by robots and computers. The total cost was estimated at $700 million, with an expected launch date of late 1991 (three years earlier than President Reagan's "within-a-decade" deadline for the space station).

The following year, NASA signed an agreement with SII whereby the agency would provide the company with three shuttle flights on credit. Once SII began generating income, it would pay NASA 12 percent of its gross revenues until the original transportation costs were paid off. It would then contract for shuttle services like any other commercial customer.

At the time, the arrangement seemed ideally suited to the new, "business-oriented" way NASA was now trying to conduct itself. Moreover, by providing support for a fledgling private company, it was also creating still more customers for the space shuttle. It is therefore hardly surprising that the project received a great deal of support inside the agency. A Microgravity Materials Science Assessment Task Force of June 1987 (chaired by Dr. Bonnie Dunbar, a shuttle astronaut), for example, noted that NASA had traditionally paid too little attention to space-based materials research. Another report, from six of NASA's Centers for the Commercial Development of Space (see last chapter), described a number of microgravity experiments they would wish to pursue on a facility like the ISF.[18]

Unfortunately, despite all of these endorsements, the venture soon ran into trouble. First, SII had seriously overestimated the short-run commercial appeal of microgravity materials processing. Although the idea of making new (or better) products in space—crystals, pharmaceuticals, metals, to name a few—has always sounded attractive in principle, the development of an actual market has been sluggish at best. The company soon reached the conclusion that the market would not "mature" until at least 1995.[19] In addition, the *Challenger* explosion and subsequent grounding of the U.S. shuttle fleet made some potential investors and users nervous: how could the company guarantee that customers would have regular access to the facility? As a result, by 1987, SII had lost one major client (McDonnell Douglas) and had managed to raise only $30 million, less than 10 percent of the amount required.[20]

In the fall of 1987, the company approached NASA again, this time with a proposal that the agency become a sort of "anchor tenant" on the ISF.[21] Under this arrangement, NASA would lease around 70 percent of the facility, at a cost of $140 million per year, for five years.[22] The federal government's involvement with the program—not to mention the five years of guaranteed income—would help the company raise the capital it needed to build the facility, and would also help reassure the company's anxious customer base.

For NASA, however, such an invitation could hardly have come at a worse time. Although its space station program had received a strong endorsement from President Reagan in 1984, members of Congress, particularly some of the more influential committee chairs, were far more skeptical.[23] Ever since it had first been proposed, for example, Representative Edward Boland, chair of the House Appropriations Subcommittee that oversaw the NASA budget, had pushed for a (relatively) smaller and simpler facility than the agency was proposing. Arguing (correctly, as it turned out) that the permanently occupied station would cost far more than $8–9 billion, as the president had claimed, Boland claimed that automated platforms and a "man-tended" laboratory could perform 80 percent of the missions proposed for the full station at 15–20 percent of its cost.[24]

Against this backdrop, NASA officials were not enthusiastic (to say the least) about a private company touting a smaller and cheaper version of one of its largest programs. They were right to be concerned: Congress was just then considering the agency's FY 1988 budget request, which asked for $767 million for the space station, more than the total estimated cost of the ISF. Not surprisingly, a number of station opponents, including Boland, Congressman Bill Green (the ranking Republican on Boland's subcommittee), and Senator William Proxmire (a long-standing critic of expensive space projects), found the facility very attractive indeed. Thus, in the FY 1988 continuing budget resolution, Congress ordered that $25 million of NASA's station budget (which had already been cut nearly in half from that requested, to $387 million) be set aside as a "placeholder" on the ISF. The resolution also directed NASA "to conclude a satisfactory funding arrangement that will lead to a workable leased ISF vehicle in the 1991/1992 timeframe [sic]."[25]

NASA, as might be expected, vigorously opposed this move. Although few in the agency would say so publicly, it was widely reported that the biggest fear was that ISF would ultimately weaken political support for the larger space station.[26] Officially, however, the agency took the position that it had no "identified needs that would justify a major commitment to use the ISF capability."[27] Ironically, for a time it appeared as though this line of argument would produce exactly the effect NASA wished to avoid: many saw it as undercutting the argument for the space station.

In an angry joint response to NASA's "no identified needs" statement, for example, Senator Proxmire, along with Congressmen Boland and Green, noted:

> Such a statement seems to suggest that . . . NASA has no requirements for microgravity research—and that causes us to question the purpose of building a space station.[28]

To make sure that it took the ISF proposal seriously, the congressmen state their intention to withhold $90 million of station development funding "until these issues are settled."[28]

One especially interesting point about this letter is its observation that "[o]ften it appears that there is an element within NASA which is only interested in building a permanently-manned [sic] space station," without regard for whether this "results in a space station that has little or no resources available for actual experiments."[30] Thus, like Representative Zschau's claim four years earlier about NASA's "eagerness" to find justifications for the shuttle, the congressmen appeared to believe (again, not without reason) that the agency was pursuing an agenda of its own.

While Congress' support for the private facility appears to have been based on economics, many senior administration officials tended to view the matter somewhat more ideologically. In language that would fit perfectly with any privatization initiative, for example, Commerce Secretary William Verity expressed his strong support for leasing space on the ISF:

> By buying services, the government will make the private sector responsible for design, financing, development, and operation. This reform will save the government money, reduce the amount of funding it must supply up-front and shift the risk of cost overruns to the private sector.[31]

Gregg Fawkes, director of DOC's Office of Commercial Space, was more adamant, as well as more directly critical of NASA. As he saw it, the agency was simply trying to maintain its "monopoly control" of space.[32]

Verity soon brought the ISF issue to the White House Economic Policy Council, one of the Reagan-era cross-cabinet organizations described in the preceding chapter. He was able to persuade other council members, such as Treasury Secretary James Baker and Transportation Secretary James Burnley to support the anchor-tenant proposal. As he had attempted with the Congress, NASA Administrator James Fletcher argued that the agency had "no requirements for the particular capability that ISF is proposing."[33] On January 7, 1988, however, the council voted to recommend to the president that NASA make a commitment to SII to lease space on the ISF.[34]

Thus, the administration's new National Space Policy, issued on February 11 (see below), announced that the federal government would indeed become the "'anchor tenant' in an orbiting research facility suitable for research and commercial manufacturing that is financed, constructed, and operated by the private sector." The contract was to be awarded that summer, with "services available to the government no later then the end of 1993."[35]

NASA was, however, able to gain two concessions. First, unlike the $25 million "down payment" of the previous year, the lease money was not to come out of its budget (the president's directive called for the facility to be used by "various Federal agencies with interest in microgravity research"[36]). Second, the agency convinced the administration that the choice of company ought to be open to competitive bidding, rather than simply issue a sole-source contract to SII. The name Industrial Space Facility was dropped, and from this point forward the project became known as the Commercially Developed Space Facility (CDSF). One month after the president's announcement, the administration had drafted a request for proposals and was "ready to get the process under way."³⁷

This, as it turned out, was the program's high-water mark. Despite its earlier endorsement by a few of their colleagues, some senators—most notably Ernest Hollings, chairman of the Committee on Commerce, Science, and Transportation, and Donald Riegle, chairman of the Subcommittee on Science, Technology, and Space—had serious misgivings about CDSF and the process surrounding it. They urged delaying release of the request for proposals until a more thorough study of the whole project could be made.³⁸ Although more sharply divided, the House Subcommittee on Space and Aeronautics also had concerns.³⁹ In May, four senators—Hollings and Riegle, joined by ranking Republican members John C. Danforth and Larry Pressler—ordered that the project be delayed pending an independent study by the National Academy of Sciences.⁴⁰ That report, released in March 1989, concluded that there would be no real need for such a facility until at least the late 1990s, by which time, it was assumed, the space station would be completed.⁴¹ Combined with the change in administration (George H. W. Bush had never commented about the project), this effectively killed the CDSF.

Can NASA Run A Business?

It is, of course, possible to argue (as at least some observers did at the time⁴²), that by requiring NASA and other federal agencies to lease space on the ISF/CDSF, the administration was violating its own (often-stated) restriction on public subsidies to private space companies.⁴³ Nevertheless, the most common descriptions of the episode were that of NASA opposing the small start-up company, fighting to block SII's project primarily in order to preserve its big-budget project. Combined with the earlier controversy over STS pricing, an image was beginning to emerge (fairly or not) that the agency was fundamentally "antibusiness."

Even in instances where NASA was sincerely (from its own perspective) trying to help the commercial space sector. A 1989 IEEE report, for example, accused the Office of Commercial Programs of "meddling" in matters that ought to be left solely to private business, and of viewing itself (rather than the financial or capital markets) as the "fundamental 'economic filter' in the space commercialization decision-making process."⁴⁴ According to the report, the office's attitudes raised

> [t]he disturbing question of whether or not NASA really understands how entrepreneurial elements of the United States free enterprise system actually go about the creation of new private sector businesses, and the truly useful, but sharply limited, role which the federal government can play therein.⁴⁵

A similar sentiment was expressed by DOC's Gregg Fawkes, who during the ISF debate commented that NASA did not "understand what 'commercial' means in space, and wouldn't like it if they did."⁴⁶

This assessment of the agency and its proper role in the area of space commerce is clearly evident in the revised National Space Policy. Like its predecessors, the document reaffirms U.S. government support for a "separate, non-governmental commercial sector."[47] This time, however, the language is more precise and the tone a bit more pointed:

> [T]he United States government shall not preclude or deter the continuing development of a separate, non-governmental Commercial Space Sector. . . . Governmental Space Sectors [sic] shall purchase commercially available space goods and services to the fullest extent feasible, and shall not conduct activities with potential commercial application that preclude or deter Commercial Sector space activities except for national security or public safety reasons. Commercial Sector space activities will be supervised and regulated only to the extent required by law, national security, international obligations, and public safety.[48]

The business of NASA, it seems, was not business.

Technical Problems

At the same time this discussion was taking place, the agency was struggling with an unusual number of highly visible technical setbacks, ranging from the merely annoying to the catastrophic. Taken together, these mishaps and failures not only severely damaged NASA's reputation but were, in at least one case, directly responsible for a major and far-reaching change in U.S. space policy.

Challenger and Its Aftermath

Obviously, the most serious of these events was the loss of the space shuttle *Challenger* on January 28, 1986. Just 73 seconds after liftoff, the spacecraft exploded in a massive ball of flame, completely destroying the orbiter and killing all seven astronauts. Subsequent investigation by NASA and an external commission revealed that the accident had been caused by the failure of an O-shaped gasket meant to seal the joints between the different sections of the solid rocket boosters that help lift the shuttle into orbit. The chilly January weather (temperature at liftoff was only 36 degrees Fahrenheit, and during the previous evening had dropped into the high 20s) had partially frozen the rubber gasket, leaving it unable to expand and form a proper seal. The escaping hot gasses ignited the fuel in the shuttle's external tank, resulting in the explosion.[49]

Amid the shock and grief, numerous accusations and recriminations were heaped on NASA. Even before the first shuttle flight, critics had argued that the Space Transportation System was technically suspect, economically inefficient, and perhaps even downright dangerous.[50] The successful launch of *Columbia* in 1981 slowed, but did not fully stop, the questions and criticism.[51] The general view of the shuttle's detractors was that, for purely political reasons (create work for itself, continue the uneconomical human spaceflight program, etc.), NASA had committed itself—and, since STS was to serve as the only U.S. launch system, the entire nation—to a fatally flawed technology.[52] To many, *Challenger* was the tangible—and tragic—proof of what they had been saying all along.

Interestingly, it did not appear that NASA suffered a loss of political support because of the accident—at least not directly. President Reagan, for one, continued to speak of the agency (shuttle and all) in typically lofty terms. In fact, there were times that he even made it appear that accidents like *Challenger* were only to be expected, a normal and accepted part of exploring space:

> They [astronauts] know that exploration has its risks. They know that with adventure, goes danger. They all know this, but they also know something far more important: something about the spirit and sense of joy that have kept man reaching through the ages to grasp for the limits of his universe and beyond that, despite hardships and peril.[53]

Perhaps more significant was the fact that Reagan still supported NASA, at least up to a point. Seven months after the accident, the president declared:

> NASA and the shuttles will continue to lead the way, breaking new ground, pioneering new technology, pushing back the frontiers. . . . NASA will keep America on the leading edge of change.[54]

Moreover, this support seemed to translate into actual policy: the administration continued to support the space station,[55] and it approved the purchase of an orbiter to replace *Challenger* (the *Endeavor*).

Nevertheless, the accident did have a major impact on the agency, one far greater than was generally recognized at the time. First, it accomplished what all of the arguments from the Office of Management and Budget, the Department of Transportation, and several members of Congress could not: it got NASA out of the commercial launch business. In October 1986, the president issued an order that, with a few minor exceptions, "NASA shall no longer provide launch services for commercial and foreign payloads."[56]

This new policy, which in effect invalidated NSDDs 42, 144, and 181, was almost certainly directly responsible for the rapid growth of the commercial launch industry. With NASA—and its subsidized shuttle rates—out of the way, the number of private launches grew rapidly, from 2 in 1989 to 22 in 1998 (for a total of 106 over the 9-year period).[57] Total revenues for the industry were projected to reach $1.3 billion in 1990. Moreover, NASA's prediction that the removal of the shuttle from the marketplace would benefit Arianne more than U.S. industry turned out to be incorrect, at least in the long run. In 1998, for example, American companies conducted 47 percent of all commercial launches worldwide, compared to Arianespace's 25 percent (Russia was third with 14 percent).[58]

Clearly, the growth of this industry was also aided by another of the administration's post-*Challenger* decisions. The 1988 Space Policy directed all federal agencies to "procure existing and future required expendable launch services directly from the private sector to the fullest extent possible."[59] Within a few years, virtually all government satellite launches, including most of NASA's (and even DOD's),[60] were being conducted by private companies. In effect, this decision changed the federal government from the industry's biggest competitor into one of its major clients.[61] Thus, while no one would ever regard *Challenger* as anything less than a terrible national tragedy, it did represent a sort of victory—after their apparent defeat with NSDD 181—for the pro-business segment of the space community, moving U.S. space policy precisely in the direction they had advocated throughout 1983 and 1984.[62]

Second, the accident—and the subsequent grounding of the entire shuttle fleet—vividly demonstrated the danger of relying on a single launch system. The Department of Defense was particularly alarmed at the prospect of having no access to space for an extended period. Accordingly, the National Space Policy recognized the need for "U.S. space transportation systems that provide sufficient resiliency to allow continued operation, despite the failure of a single system."[63] Thus, in a move somewhat parallel to the rise of the commercial launch sector, the Pentagon began once again to procure and operate expendable launch vehicles.[64] Even after the shuttle began operating again in 1988, major military payloads (which would probably still have been permitted to use the system even under the new guidelines) virtually disappeared from its manifest.[65]

In short, by 1988, STS had lost all of its business and military customers. Despite its return to flight status, the addition of a replacement orbiter, and even the president's strong endorsement, the largest NASA program since Apollo had no assigned role to play in either of the nation's primary space activities. Combined with the removal of commercial activity from the space station (see below),[66] NASA was finding itself increasingly removed from the new direction the space program was taking.

Spiraling Costs, Dwindling Performance

Even before the *Challenger* accident, it was clear that STS could not live up to NASA's ambitious vision of a decade earlier. The agency had predicted at least 2 shuttle flights per month by the mid-1980s; in 1985, its last full year of operation (pre-*Challenger*), there were only 6, and only 24 total over the 5-year period.[67] During the 1970s, NASA had promised that shuttle flights would be economical, costing as little as $400 per pound. By the mid-1980s, however, some estimates put the costs of a launch at more than $400 million, or around $10,000 per pound.[68]

Similar problems confronted the space station. At the time Reagan approved the program, NASA had estimated that the facility would cost around $8 billion over a decade.[69] By 1990, after a seemingly endless series of redesigns and "rescopings,"[70] that figure had risen to $30 billion, and some sources were predicting that the total price (including operating costs) would reach as high as $120 billion.[71] Moreover, even as it was becoming more expensive, the station's capabilities were being scaled back. Originally envisioned as a series of laboratories supporting a wide range of scientific and commercial activity (astronomical and earth observation, research in biology and medicine, materials processing), as well as (eventually) a docking port for further human expeditions into space, by the end of the 1980s, NASA had steadily whittled its objectives down to one, biomedical research.[72] In effect, critics claimed, astronauts would be permanently stationed in space simply to make observations about themselves.[73]

The selection of the station's primary mission is significant. Chapter 7 noted that although NASA had been touting the commercial value of a space station, its primary objective had been to create a permanent human habitat in space (hence its continued characterization of the project as "the next logical step"[74]). Such a facility was necessary, many space enthusiasts felt, to prepare for long-duration missions like a human expedition to Mars.[75] If this could be done in a way that also accomplished the administration's objectives, so much the better. Rising costs and mounting technical problems, however, required that some of the station's functions be scrapped, and it is hardly surprising that NASA chose to preserve the one relating to its highest priority. That decision, of course, did little to endear the project to policymakers (as seen, for example, in the debate over ISF).

Cost increases and technical failures were by no means limited to these two programs. NASA was highly embarrassed over the misaligned mirrors of its $2 billion Hubble Space Telescope[76] and the total loss of the $1 billion *Mars Observer*.[77] Although neither was a commercial program (both were sponsored by the Office of Space Science), their problems clearly contributed to the growing image of NASA as an agency in disarray. As one critic put it, the nation's premier space organization was beginning to look "silly."[78]

The problems of this period, however, go beyond simple embarrassment. The types (and the frequency) of technical malfunctions and delays the agency was experiencing during the 1980s were clearly figuring into the private sector's investment decisions. As noted earlier, for example, the *Challenger* accident made it far more difficult for Space Industries Incorporated to raise start-up funding for the ISF. A 1989 Congressional Budget Office report notes:

> The private sector's return on its investment will inevitably depend upon having assured access to the public infrastructure. This access is not only constrained by government policy, but also by events unaffected by policy such as technical delays or catastrophes like the Challenger [*sic*] accident. If the recent history of NASA infrastructure efforts is any guide, the schedule for its new initiatives is likely to experience delays similar to those of the past.[79]

In other words, setbacks like Hubble and the *Mars Observer* were simply one more reason why NASA needed to be kept far away from commercial operations.

A New Definition, A New World

Of course, it is possible to argue that there is a bit of a double standard operating here. During the Apollo era, NASA certainly had its share of mishaps, failures, and even fatalities. In fact, when comparing the success rates of earth-orbit launches during the 1960s (85 percent) with those of the 1980s (93 percent), it appears that the agency was actually performing better than it had during its "glory days."[80] Moreover, when adjusted for inflation, NASA was spending far less money during the 1980s ($109 billion) than it had 20 years earlier ($157 billion).[81] Such comparisons, however, only serve to underscore the changes resulting from the new way space policy was defined.

As chapter 2 noted, different "types" of public policy are often evaluated according to different criteria. Thus, when considered as a sort of national security issue, as it was during the height of the Cold War, political leaders and the public at large were far more willing to accept high costs and relatively higher risks from the space program. After all, as seen in chapter 4, some politicians were characterizing projects like Mercury and Apollo as matters of "national survival" during the late 1950s and early 1960s.

One minor—albeit telling—episode of that era involves the treatment of the chimpanzees that rode the first Mercury capsules into space. Writer Tom Wolfe has suggested that some of these animals were actually abused if they did not cooperate with researchers (in some cases, they were even beaten with rubber hoses). It is unimaginable that such treatment would be tolerated today; indeed, under current guidelines governing animal experimentation it would clearly be illegal. At the time, however, it was seen as a perhaps regrettable (Wolfe states that the researchers did not see themselves as being intentionally cruel) but necessary part of protecting the national interest.[82]

Throughout the 1980s, civil space policy became increasingly identified with commercial operations. This had the effect, first, of placing the program overall somewhat lower down on the scale of national priorities. Although space exploration did not become unimportant during this period, it was, in the administration's view, less important: making money is generally regarded as less critical than "national survival" (and, as seen in the last chapter, Reagan was spending far more money on military space applications).

A second, and somewhat related, result of this change was that NASA began to be held to a standard that placed far more emphasis on such factors as cost and efficiency than it was during the 1960s. Policymakers and the general public were more willing to "spare no expense" in the drive to establish supremacy over the Soviets, an idea that Eisenhower feared and tried unsuccessfully to avoid. In addition, during the early days of spaceflight, when much of the relevant technology was still under development, public officials were more understanding of technical mishaps and accidents.[83] Toward the end of Reagan's second term, however, when spaceflight seemed to have matured to the point where it was even possible for private businesses to engage in it, technical failures on the part of the government program (notwithstanding the fact that its objectives were often far more challenging) were viewed much more critically.

Finally, by the end of the 1980s, NASA was operating in an environment where its presence—if not its very existence—was increasingly suspect. Under the Reagan administration's new policy definition, the role of government had been changed from that of a direct service provider to more of a "facilitator,"[84] in which the primary goal was to help others move into space. As chapter 2 suggested, this new role required a rather different type of organizational structure. For nearly 20 years, NASA—which had been established (and organized) according to the Cold War definition of space policy—had been struggling to carry out a new, service-based mission. The Reagan administration, with the support of the business community and free market conservatives generally, had chosen the somewhat more expedient path of simply creating new agencies, such as those in DOT and DOC, to carry out its commercial initiatives. Thus, by the end of Reagan's second term, a very large question loomed for space advocates: What was NASA's purpose?

NASA at the Margin

For the most part, the transformation of space policy into an economic issue continued—and, in some respects, even accelerated during the presidency of George H. W. Bush. Ten months after taking office, the administration issued its own official statement on space, National Service Directive (NSD) 30, which in many respects was mirrored Reagan's revised Space Policy of January 1988. One of the major changes was purely organizational: a new National Aeronautics and Space Council, chaired by Vice President Dan Quayle, replaced the Reagan administration's SIG (Space).[85]

Any resemblance between this organization and that headed by Lyndon Johnson during the early 1960s, however, was limited to the name. Its membership was more diverse then the defense-heavy SIG (Space), consisting of the secretaries of state, defense, treasury, commerce, and transportation, as well as the directors of the OMB and the CIA, the president's chief of staff and science advisor, and the NASA administrator. Nevertheless, under Quayle, an economic (and social) conservative, the Space Council generally adopted a strongly pro-business stance, on occasion even going so far as to seek industry advice (and thus bypassing NASA) on matters of space policy.[86]

To be sure, Bush—like Reagan—was a staunch supporter of the space station, despite the fact that it was no longer viewed as a commercial facility. Unfortunately for NASA, the project was facing increasing opposition in Congress, and was saved from outright cancellation on a number of occasions only by direct White House lobbying.[87] Given the constant struggle to save its only major program, there was usually little opportunity to address the larger question of the agency's overall mission.

Space Exploration Initiative: The Exception that Proves the Rule

For a few months in 1989 and 1990, however, it appeared that NASA's fortunes were about to improve. In a speech at the National Air and Space Museum marking the 20th anniversary of first moon landing—and with the *Apollo 11* astronauts at his side, President Bush unveiled his Space Exploration Initiative (SEI), a bold plan to return the nation to the moon and, eventually, to send humans to Mars.[88] Under this proposal, the United States would set up a permanent research facility on the moon by 2000 and, using this base and the space station, would launch an expedition to Mars sometime between 2010 and 2019. The total cost of the venture was estimated at $500 billion over 20 years.[89]

It is worth noting that, while Bush's National Space Policy, issued shortly after the museum address, does indeed call for a "long-range goal of expanding human presence and activity beyond Earth orbit and into the solar system" and promises a forthcoming "presidential decision on a focused program of manned exploration of the solar system,"[90] this language actually comes, literally word for word, from Ronald Reagan's Space Policy of 1988.[91] Unlike Reagan, however, who did not follow up this rather broad language (which one observer has characterized as a "mild gesture" to NASA's Office of Exploration[92]) with any specific proposals, Bush was ready to be explicit. At the 1990 commencement ceremony at Texas A&M University, Bush declared:

> Leadership in space takes more than just dollars. It also takes a decision. And so, I'm announcing one today. . . . I believe that before Apollo celebrates the 50th anniversary of its landing on the moon, the American flag should be planted on Mars.[93]

To space advocates, this was a dream come true. It had long been assumed in this community (at NASA and elsewhere) that the only reason that the United States had not had a "grand vision" for space after Project Apollo was the lack of just such a declaration, another Apollo-like commitment.[94] At long last, it seemed, the wait for the required "presidential leadership" had finally arrived. Moreover, unlike rocket launches or (according to ISF's supporters) a space station, a lunar base and a Mars expedition were unquestionably projects that only NASA—not some other government agency or private company—could perform. Thus, SEI appeared to present the agency with a larger purpose, a mission, that moved it out of the supporting (or, as some would have it, obstructionist) roles it had been playing for the past few years.

Sadly for the enthusiasts, the dream faded quickly. Looking back, it seems safe to say that one of the more remarkable aspects of the SEI experience was the general indifference to the president's announcement. Particularly when compared to Kennedy's "moon speech" or Reagan's call for a space station "within a decade"—both of which it was clearly attempting to emulate[95]—Bush's proclamation seemed to hit the country with a resounding thud.

This is not to say that the proposal was ignored altogether. Congressional opponents immediately attacked it as a wasteful extravagance at a time of extreme budget stringency.[96] Even the criticism, however, seemed somewhat muted. The fact is that, although it was discussed in Congress and elsewhere,[97] it almost appears as though few Americans took SEI very seriously. Unlike the stormy debates over other big-budget R&D programs of the period, such as the Superconducting Super Collider (or, for that matter, the space station), congressional action on the moon–Mars project, took on a rather detached air. Its death in the summer of 1990 was not marked by rancorous speeches or a closely watched final confrontation on the floor of Congress (again, the demise of the Super Collider makes a telling contrast[98]), but rather by a sort of quiet fade-out.[99]

The analysis presented in chapter 2 suggests that the primary reason the Space Exploration Initiative possessed such an "unreal" quality was that it fell outside of any accepted definition of space policy. Kennedy's lunar landing goal was cast—and widely understood—as a direct challenge to a feared adversary. Similarly, the space station—at least as it was originally conceived—fit (albeit imperfectly) into the Reagan administration's larger vision of encouraging space commerce (as well as Reagan's own enthusiastic vision of America's destiny). Even the space shuttle, which was not unveiled in a particularly dramatic fashion, was described in terms that conformed—if somewhat vaguely (see chapter 6)—to some broadly accepted notion of the space program's purpose.

This was clearly not the case with SEI. There does not appear to have been any sort of effort to link it to national security policy (and it is difficult to imagine any public official regarding such a connection as credible). Bush and Quayle both referred to it as "an investment in America's future," but other administration officials, such as Science Advisor D. Alan Bromley, specifically rejected classifying it as an economic or even an R&D program:

> Some . . . projects [like] the Space Exploration Initiative . . . have never been primarily science projects, although for convenience they are included with R&D in the budget. But they are justified primarily on the basis of considerations such as exploring physical frontiers, increasing appreciation of science and technology, maintaining national leadership in a given area, or taking the initial steps in a great human adventure and only secondarily on their long-term (though quite certain) contributions to scientific and technological progress. To criticize them on the grounds that they do not advance science sufficiently is to fail to recognize their multiple objectives.[100]

One of the more striking features of this statement is how the reference to SEI's "multiple objectives"—none of them stated specifically—harkens back to the "multiple capabilities" (but no clear mission) description of the space shuttle during the 1970s (see chapter 6).

Finally, unlike Kennedy, Johnson, or Reagan, whose space proposals were shaped (for the most part) by their administrations' respective larger political and ideological objectives, Bush's SEI did not readily lend itself to any sort of overall goal that he might have set for his presidency.[101] In short, lacking any generally recognized policy definition that could conceivably cover it, the Space Exploration Initiative came across almost as an irrelevance, an idea that simply did not "fit" American political discourse of the late 1980s and early 1990s.

The Goldin Age: The End of Business as Usual?

Although the Space Exploration Initiative, for all of its initial fanfare, ultimately had little impact on the program, another of Bush's decisions was to prove somewhat more long-lasting. By the early 1990s, it had become clear that NASA's organizational structure and procedures represented a serious impediment to its full participation in U.S. space policy as it was now defined (although obviously no one used the term at the time). Accordingly, in 1992 the president appointed a new administrator, Daniel Goldin, who promised to "shake up" the agency.

A former executive with the TRW corporation (as well as a former NASA employee), Goldin immediately set out to change what he called the "organizational culture" at NASA, largely by forcing it to adopt more of the practices—and the attitudes—of private industry. He pushed, for example, for a greater reliance on newer, more "cutting-edge" (and lower-cost) technologies. He sought to introduce the principles of total quality management (widely employed by many firms at the time) into the organization, going so far as to appoint an associate administrator for "continuous improvement." Finally, he moved the agency away from very large, so-called behemoth space missions like the *Mars Observer* and the *Galileo* Jupiter and *Cassini* Saturn probes,[102] opting instead for a larger number of smaller, more focused projects. The phrase *faster, better, cheaper* was to become a sort of mantra within the agency.

Organizationally, NASA under Goldin began to contract out a greater share of its work than ever before, including, ultimately, day-to-day operations of the space shuttle. This, in turn, led to a significant downsizing of its permanent staff. The number of employees (excluding contractor personnel) fell from approximately 25,000 in FY 1993 to 20,000 in FY 1995.[103] Efforts to consolidate or even close some of the agency's field centers, however, were forestalled by heavy congressional opposition. As with any major facility (such as a military base), which usually represent significant sources of economic activity and employment in individual congressional districts, elected officials almost always act quickly to protect these facilities.

The political controversy over the field centers underscores a serious problem with respect to Goldin's reforms, namely, that he was not—and, under the American political system, could not be—completely in control of his agenda. It is significant, for example, that almost all of his technical initiatives relate to space science missions, which generally do not receive much legislative or presidential scrutiny. Even if he had been inclined to make any major decisions on NASA's larger projects, such as the shuttle and the space station (which, not coincidentally, are associated with human space flight), he would almost certainly have had to seek some sort of approval from the administration or Congress.

To make matters worse, an agency like NASA must face other, nonpolitical limitations as well. Throughout 1999, it encountered a series of technical problems, some of which raised serious questions about Goldin's "faster, better, cheaper" approach. First, in the beginning of the summer, the entire shuttle fleet was grounded due to faulty wiring. Then, in June, the *Mars Climate Orbiter* was lost just as it was about to enter Martian orbit. Subsequent investigation revealed that the problem resulted from a failure to convert all of the metric measurements in the *Orbiter*'s navigation data.[104] Finally, in December, the Mars Polar Lander crashed on the Martian surface due to what was believed to be a problem in the spacecraft's software.[105] Critics began to point out that the failures of the probe reveal the deficiency of relying so heavily on lower-cost, but largely untried, technology.

Overall, Goldin's tenure at NASA clearly demonstrates the limits of "internal reform" as an answer to the agency's problems. In particular, no amount of internal restructuring, even if successful, can by itself really address the question of NASA's purpose. As James Webb noted a generation earlier (see chapter 6), that answer must come from higher up.

Political Change, Policy Inertia

In politics, the clearest indicator that a major policy shift has acquired some degree of permanence is how well it endures from one presidential administration to the next, particularly when that transition involves a change in political party. As a fellow Republican, it is hardly surprising that Bush retained much of Reagan's overall approach to space policy. What is surprising is the extent to which that approach was accepted, and even furthered, by Bush's successor.

Despite the fact that Democratic president Bill Clinton was far less hostile to government programs—and, for the most part, less overtly "pro-business"—than either of his predecessors, his administration's official declarations on space continued to place a high priority on business and commerce, often employing language virtually identical to that of Reagan and Bush.

Clinton's 1996 National Space Policy, for example, noted that "expanding U.S. Commercial Space Activities will generate economic benefits for the Nation," and provide "an increasing range of space goods and services." It called for the identification and elimination of "laws and regulations that unnecessarily impede commercial space activities." As with similar Reagan and Bush documents, Clinton's Space Policy also prohibits "use of direct Federal subsidies."[106]

In his 1995 Space Transportation Policy, however, the president took the privatization of space one step further. The primary purpose of this document was to begin the process of developing a successor to the space shuttle, which by that time had been in service for nearly 15 years. Some parts of this directive simply reaffirm such post-*Challenger* practices as limiting use of the space shuttle to missions requiring "unique Shuttle capabilities," directing the government to purchase commercial launch services "to the fullest extent feasible," and proscribing practices "that preclude or deter commercial space activities."[107] The directive parts company with similar documents from the Reagan and Bush years, however, in its approach to developing a new launch vehicle:

> It is envisioned that the private sector could have a significant role in managing the development and operation of a new reusable space transportation system. In anticipation of this role, NASA shall actively involve the private sector in planning and evaluating its launch technology activities.[108]

Finally, in a remarkable departure from past practice, the directive calls for the federal government to encourage private sector *financing* of any new launch system.[109]

This move toward still greater reliance on industry reflects not only the ongoing change in policy (see below), but was also a testimony to the growing dominance of the commercial space sector. In 1999, the total amount spent on space activities by private firms (worldwide) exceeded total government expenditures for the first time ever. Moreover, public budgets for space were declining in every space-faring country except India.[110] Clearly, a major transition in the conduct of global space policy was taking place.

Of course, proponents of Reagan/Bush–era space policy might well argue that increasing business' role in space is such an inherently good idea that even a Democrat like Clinton had to accept it. The more likely explanation, however, is that Clinton's approach to the program represented a passive acceptance of his predecessors policies, rather than an active endorsement. Unlike Reagan or Bush, Clinton seemed to take little interest in the space program. He made relatively few public statements about space during his eight years in office, and except for his National Space Transportation Policy,[111] put forward no proposals for any new projects. By 1998, NASA's budget was nearly $3 billion less than when Clinton took office (interestingly, space spending in the Department of Commerce went up by more than 20 percent over the same period).[112]

The most revealing factor, however, was the way in which the Clinton White House organized its space policy. Rather than creating a central coordinating body like SIG (Space) or the National Aeronautics and Space Council, the Clinton administration relegated space decision making to the National Science and Technology Council, a department of the White House Office of Science and Technology Policy. This had the effect of not only moving space policy somewhat lower in the hierarchy of the executive branch, but—since council's role is to coordinate all federal R&D policies—of also ensuring that it would receive less special attention than before.

Ironically, all of this was occurring in an administration that actively sought to involve the federal government in technological development generally. By the time Clinton entered office, the practice of public–private R&D collaboration—primarily by means of formal cooperative research and development agreements (CRADAs)—was already well established.[113] In keeping with the conservative view about the role of government (see previous chapter), however, these had been limited to so-called precompetitive technologies or (more often) to those with direct defense application. Any commercial applications that might "spin off" from this activity were welcomed, but they were never considered to be its primary purpose.[114]

From the beginning, Clinton officials took a very different view of the federal role in R&D. One month after taking office, the administration declared:

> We cannot rely on the serendipitous application of defense technology to the private sector. We must aim directly at these new challenges and focus our efforts on the new opportunities before us, recognizing that government can play a key role helping private firms develop and profit from innovation.[115]

Not surprisingly, the number of CRADAs—in areas ranging from automobiles to computers to textiles—jumped from 29 in 1991 to 90 in 1993.[116] In addition, Clinton boosted funding for the Advanced Technology Program (which had actually been created during the Bush administration), an agency in the department that provides federal funding for private companies pursuing particularly expensive or high-risk R&D.[117]

The fact that such a "pro-technology" president would (relatively speaking) neglect NASA and its programs suggests that he was not inclined to define space as a technology issue. Whereas Reagan officials identified space policy as an integral part of their larger political aims (particularly with regard to economic privatization and national defense) early on, the Clinton administration never seemed to make such a connection.

New Actors and a (Small) Redefinition

This does not mean, however, that the program was neglected entirely. In 1994, the Republican Party captured control of the U.S. House of Representatives for the first time in 40 years.[118] The new chairs of the relevant committees and subcommittees—most notably Robert Walker, chair of the Science Committee, and James Sensenbrenner, chair of the Subcommittee on Space and Aeronautics—were staunch supporters of the space commercialization initiatives begun during the Reagan administration.[119] Their enthusiasm, combined with the relative indifference of the White House, meant that, by the end of the decade, Congress had become more influential in space policy than any time since the late 1950s and early 1960s.

Still, to a limited degree, the Clinton administration was responsible for a minor redefinition of one part of the space program. In late 1993, the United States invited the new Russian government to participate in the space station program—renamed the International Space Station—joining Japan, Canada, and the members of the European Space Agency. Ostensibly, this was done in order to take advantage of the vast experience in station operations and long-duration space missions that developed in the Soviet program.[120] It soon became clear, however, that the administration had an additional reason for supporting Russian involvement.

The transition to political democracy and a market economy in the former USSR was not going well. The substantial economic dislocations brought about through the dismantling of the state planning apparatus and state ownership of major industries (which were already far less efficient than their Western counterparts) had imposed severe financial hardship on much of the population. Not surprisingly, the government found that it could not possibly afford the sort of space program that had enjoyed such a highly privileged position under the Soviet system.[121] Under the arrangement with NASA, Russian firms were to build and launch several of the International Space Station modules, paid for in part (in large part, as it was to turn out) by the United States. From the administration's point of view, this would allow the United States to support the Russian economic transition (a sort of "back door" foreign aid) and keep its scientists and engineers gainfully employed (as opposed to offering their services to a terrorist organization or a country like North Korea or Iraq). In other words, Clinton officials had come to view (and, in effect, redefine) the International Space Station as an instrument of their foreign policy.

Whether this approach will succeed in revitalizing the Russian economy remains to be seen (the results to date have not been encouraging[122]). As a *space* policy, however, the administration's strategy had significant drawbacks. Already seriously behind schedule (Reagan had originally called for it to be operational "within a decade," that is, by 1994), the station program was thrown even further behind, largely due to the Russian involvement. The delays proved not only costly (subsidies to the Russian Space Agency greatly exceeded the initial estimates), but in at least one case posed a serious threat to the project itself.

By early 1999, the Russian-built *Zvezda* module—a vital component which, in addition to providing habitation space for the crew was to be the station's primary source of propulsion and altitude control—was still not ready for launch.[123] Without this module, there would be no way to keep the station elements launched the previous year in their proper orbit. NASA was forced to ask Congress for additional funding to develop an interim control module as a temporary substitute. Needless to say, the Republican congressional leadership, which had long been critical of the Clinton approach, was not at all pleased.[124]

This, once again, underscores the problems that arise when policymakers adopt differing definitions of an issue. Viewing Russian participation as a foreign relations problem, the administration seems more willing to accept the delays (and, since the conduct of foreign affairs is generally seen as a presidential prerogative, are less inclined to accept Congressional "interference") than the Space and Aeronautics Subcommittee, which, naturally enough, sees it as matter of space policy. It seems quite unlikely that this conflict will be resolved anytime soon.

Conclusion

By the end of the 1990s, there were clear signs that the sense of direction imparted to U.S. space efforts by the redefinition 15 years earlier was beginning to unravel. The first indication of this was the ill-fated Space Exploration Initiative, which raised serious questions as to how President Bush was defining space policy (or if he even had such a definition in mind). Matters did not improve during the 1990s, with a president whose vision of the program—to the extent that he had one—differed considerably from that of the congressional leadership. Meanwhile, NASA did little to help itself, ending the decade with a series of highly visible (and expensive) technical failures.

The fundamental problem here, which had been quietly smoldering since Reagan's second term, was the growing separation between space policy and NASA policy. As chapter 7 noted, the dominant thrust of the Reagan redefinition—defense and commercialization—left no clear role for NASA. This fact had been partially obscured by proposals like the space station, which seemed, at least initially, to fit into the prevailing definition, while at the same time providing substantial work for the space agency.

As the 1990s drew to a close, however, NASA's position seemed increasingly precarious. Within the growing number of space policy stakeholders—executive branch officials, members of Congress, industry leaders, the scientific community, and the like—whatever consensus there had been about NASA's purpose (that is to say, its mission) was virtually nonexistent. For an agency with a multibillion dollar budget, this is not a particularly safe place to be.

9

Concluding Thoughts

The preceding analysis does not necessarily lead to a single conclusion. It does, however, provide the basis for a number of observations concerning the place of NASA within federal science and technology policy, the agency's future, and the overall status of the U.S. space program at the beginning of the 21st century. It is also possible to make one or two recommendations along the way, as well.

Defense Conversion Redux

Almost everyone involved—policymakers, scientists, members of the academic community, and the mass media—agrees that the end of the Cold War in the early 1990s represented a major turning point in the history of U.S. R&D policy. With the collapse of the Soviet Union, the major American science- and technology-based organizations were faced with two major challenges: moving beyond the traditional national security justifications for federal sponsorship of their programs, and reorienting those programs to accommodate such new demands as economic growth and U.S. industrial competitiveness. Thus, the 1990s saw a number of major initiatives to develop new missions for the national laboratories, encourage the Defense Advanced Research Projects Agency to develop more "dual-use" technologies, and generally to "convert" defense-based R&D institutions into enterprises more compatible with the civilian global economy. Along the way, these efforts have generated hundreds of books, articles, government reports, congressional hearings, and scholarly conferences, almost all of which assume (at least implicitly) that this problem is a brand new one.[1]

To NASA and its supporters, however, the issue (although not the fanfare) ought to look quite familiar. As the preceding chapters have shown, the experience of the space agency from the 1970s forward represents an example, not just of an analogous situation, but of the very same phenomenon.[2] During the late 1950s and most of the 1960s, NASA was just as deeply involved in "fighting" the Soviets as was DARPA or Los Alamos, Lawrence Livermore, or Sandia national laboratories. Unlike those institutions, however, it was (in the words of chapter 6) "retired" from the Cold War long before that conflict was over. In addition, virtually every strategy that has been proposed for (and, in some cases, enacted by) the defense establishment has already been tried by NASA sometime during the past 25 years.

In other words, nearly a quarter century before defense conversion became a "recognized" issue, NASA was itself forced to "find a new role" and to "justify its existence" on grounds other than competition with the USSR. In doing so (or, as many see it, attempting to do so), it provided an almost exact preview of what the national laboratories and others would encounter during the 1990s. That this has not even been generally recognized, let alone widely discussed, is almost as significant as the fact itself.

By 1991 (the year of the Soviet Union's collapse), the U.S. government had already spent hundreds of billions of dollars developing the largest and most sophisticated R&D infrastructure in human history. The 30 national laboratories overseen by the Department of Energy, for example, employed more than 29,000 people in 16 states, and had a combined annual budget of nearly $6 billion (they had spent more than $100 billion in just the preceding 20 years alone).[3] Their capital value was approximately $30 billion.[4]

Unfortunately (for them), their primary justification for their existence, the USSR, was now gone. Although a fair amount of activity at the labs was devoted to basic research (particularly in physics), almost half of their funding was devoted to weapons development. In fact, three of the larger laboratories—Los Alamos, Lawrence Livermore, and Sandia—existed almost exclusively as weapons laboratories. Most policymakers felt that, with the Cold War over, this was at least two too many. Even so, these were state-of-the-art facilities, staffed by some of the most brilliant scientists in the world (58 Nobel Prize winners, according to DOE[5]). Thus, the question quickly became, in the words of one science policy specialist, "How can the vitality of these laboratories be sustained and their capabilities put to the best use?"[6]

Over the following decade, there emerged two major approaches for "using" the U.S. Cold War R&D infrastructure. The more straightforward was direct "defense conversion," following the biblical adage to "beat their swords into plowshares" (or, as one newspaper article colorfully put it, "bombs into bulldozers"[7]). This was, of course, really the only option open (other than simply shutting down) to private firms that had previously survived exclusively on DOD contracts, and large numbers of companies attempted to transform their production lines from the manufacture and testing of military hardware to the production of goods for commercial sale.[8]

For the national labs, "conversion" meant, using language that will by now seem quite familiar, finding "new missions."[9] From the start, they and their supporters in Congress declared that the labs were uniquely suited to serving the needs of the civilian economy:

> The billions of dollars invested in the national laboratories have produced a host of unique tools for research, including . . . an impressive infrastructure of engineering, systems management and computational horsepower, not to mention a cadre of scientists skilled in mission-directed research. Nothing could be more natural than "leveraging" this investment to attack problems of importance to U.S. industry.[10]

Los Alamos, for example, proposed to reconstruct itself as a "research and development partner for industry" in such areas as nonpolluting cars and advanced semiconductor manufacturing, and even brought in executives from Motorola to help make its entire operation more "businesslike."[11] Similarly, in 1992 Congressman George Brown recommended turning Lawrence Livermore into a "critical technologies center," which would conduct research on such areas as materials science, biotechnology, and the next generation of computers.[12] Indeed, most of the labs began to speak of themselves as the heirs of the "cutting-edge" industrial research tradition that had once existed at IBM and AT&T.[13]

More often than not, this "aid-to-industry" function of the labs (as well as other government agencies) is conducted in partnership with one or more private firms. There are now well over one thousand such arrangements underway, involving dozens of federal facilities. Their goals have ranged from creating more economical batteries[14] to helping design the next generation of automobiles (triple fuel efficiency during the Clinton administration, hydrogen power under George W. Bush)[15] to developing high-temperature superconductivity.[16] It is still too soon to say how successful these efforts have been.[17]

The second strategy, which centered primarily on the work of DARPA,[18] was to encourage the development of so-called dual-use technologies, that is, systems that have both military and commercial applications. Although it spends as much on research and development as its main economic rivals (and, in some cases, far more than), the U.S. government devotes a far larger share of its science and technology budget to defense purposes than any other country. In 1992, for example, nearly two-thirds of U.S. federal R&D spending was defense related, compared with 18 percent in Germany and 9 percent in Japan.[19] With concerns over industrial competition rising—and the fear of the Soviet Union no longer a factor[20]—American policymakers began to reconsider how science and technology funding was being distributed. The result—(dubbed "swords and plowshares"[21]) was to emphasize work in such areas as computers (especially high-speed computing),[22] jet engines, optical fiber cable, and infrared sensors.[23]

One of the more interesting features of both of these strategies is their wide political appeal. It might be expected that, given its general stance toward technology (see the previous chapter), the Clinton administration would approve of programs such as these. What is more surprising is that many were actually initiated during the Reagan years. It was in 1986 that Congress passed (and the president signed) the Federal Technology Transfer Act,[24] that first authorized the use of cooperative agreements between private firms and federal agencies. The first Bush administration seemed even less hesitant about using the laboratories in this fashion. Secretary of Energy James Watkins, in a letter to House Science Committee Chair George Brown, noted:

> The science and technology base of the Laboratories provides what I call the infrastructure for solving large problems of great complexity. It is this infrastructure that I propose to bring to bear on the question of the competitiveness of our industries and businesses. This should be done in partnership with business and universities. . . . [B]usiness can provide the market pull on the talents of the Laboratories that will assure that their work is relevant.[25]

The approach was even endorsed by Vice President Quayle's Council on Competitiveness, which usually favored keeping government out of business decisions.[26] In short, belief in the desirability of involving the national labs in private sector development seems to transcend party and ideology.

Looking over the discussion in the last few chapters, it can be argued that NASA was actually the first government agency to employ both of these approaches. As chapter 6 pointed out, by the late 1960s the agency was in need of new justifications for its programs. Given that the most important factor driving decisions about space policy seemed to be expense, and that opponents continued to criticize program advocates for "ignoring problems here on earth," it is hardly surprising that NASA planners began about this time to speak about projects that were "cost-effective" and that promised direct economic benefits. In seeking, for the first time in its history, to employ an economic justification for its programs while still making use of the infrastructure and capabilities built up during Apollo,[27] NASA was, in effect, pursuing the very first examples of post–Cold War defense conversion, promising to use the technical expertise, administrative skills, and specialized hardware that it had originally developed to "fight" the USSR to provide goods and services for the civilian economy.

Such reasoning was used, for example, to justify the Apollo Applications (*Skylab*) Project (see chapter 6). Indeed, the project—even its title is significant— represents about as pure an example of defense conversion as one is likely to find. It utilized a great deal of the existing Apollo hardware—including the command and service modules, and a Saturn V rocket to place the *Skylab* module in orbit— in order to "help with such problems as air and water pollution, flooding, crop deterioration, and erosion."[28] The project was described by a NASA official as "one of the most significant benefit-oriented programs of the space age."[29]

Similarly, the space shuttle program can be seen as the first attempt to develop a dual-use (if not a multiple-use) technology. It was expected to serve the needs of paying customers from the private sector, as well as act as the primary

CONCLUDING THOUGHTS   169

space vehicle for the Department of Defense. Indeed, DOD had had substantial input into the system's design,[30] and the air force had even planned to construct its own shuttle launch facility in Southern California. In addition, STS was to operate as an orbital platform for basic scientific research.[31]

To be sure, NASA had "ulterior motives" in pursuing both of these projects. As noted in chapter 6, James Webb reportedly pushed *Skylab* largely because he wanted to keep the Apollo production lines open as long as possible.[32] It was also a means of keeping its human spaceflight program alive in the hope that some years down the road it might gather enough political support for still larger projects, such as a permanent space station.[33] It is difficult to see, however, how this is any different from the efforts of Livermore or Sandia 25 years later. Each organization represents a multibillion dollar taxpayer investment that, it can reasonably be argued, even if no longer needed for its original purpose, can still serve in some productive capacity.

As it turned out, however, NASA's efforts at conversion were, to say the least, rather less than successful. There was no follow-on to *Skylab*, which fell out of orbit and burned up in 1979.[34] Even had it survived, it would still have had many critics arguing that its research results did not justify its costs. The shortcomings of STS, both as a commercial and a military system, have been widely noted (not to mention the fact that two shuttles have been destroyed). By the same token, there is little evidence to suggest that the current attempts to do basically the same thing have worked (or will work) any better. Although by no means a perfect parallel, it is worth noting that a number of DOD's former contractors who sought to move into commercial manufacturing have had a very difficult time. According to Lewis Branscomb, who has written widely on the subject:

> [C]ommercial innovation is a delicate process requiring very special skills, a lot of investor courage, a high level of technical agility, and the willingness to accept failure in a significant number of cases.[35]

Having been largely shielded from the ups and downs of the marketplace by their relationship with the Pentagon, many firms found that, despite their technical expertise, they are simply unable to compete. As for the new projects underway at the national laboratories, it is (as already noted) too soon to make a valid assessment.

The basic problem here is a fundamental difference in the political situation of the 1970s and the 1990s. Although there has been some criticism of specific programs,[36] the laboratories and DARPA have, for the most part, enjoyed an enormous reservoir of political goodwill. Far from being treated as "outmoded" relics of the Cold War simply trying to hang on, they have from the start been viewed as a valuable national resource. The question was never *whether* they had a role to play in the post–Cold War world, but *what* role they should play.

Unfortunately for NASA, concepts such as "defense conversion" and "dual use" did not exist in 1970. The idea of using the complex and highly specialized (not to mention extremely expensive) hardware from Apollo to study soil erosion must have seemed, to policymakers of that era, like the worst kind of technological overkill. Thus, many politicians saw NASA's proposals as little more than an agency's far-fetched efforts to save its budget. By the 1990s, in contrast, having a large, sophisticated technical organization work on (relatively) smaller problems became a generally accepted, if not encouraged, practice.[37]

This is in no way to suggest that government officials have somehow been unfair to NASA, or that the agency is the victim of some sort of double standard. Just as sending a single unit home from a war is far different than the general demobilization that follows an armistice, there is an enormous difference—politically, economically, and socially—between the "reassignment" of one Cold War agency and the end of the Cold War itself. Put another way, the very different circumstances under which NASA and the rest of the Cold War R&D community "completed" their "missions" has given rise to very different definitions of the problem involved: the efforts of the national labs, DARPA, and other similar organizations have been defined as "defense conversion"; NASA of the 1970s, as was pointed out in chapter 6, was unable to fit into any prevailing issue definition whatsoever.

What is being suggested here is that the policymakers who are concerned with these issues might well profit from a review of the agency's experiences with conversion, dual technologies, and the like. Unless one is willing to presuppose that NASA during the 1970s and early 1980s was significantly less competent than, say, the national laboratories are now, examining the early history of space commercialization could provide a useful guide to present-day conversion efforts. If nothing else, it might help point out policies or programs to avoid.

Red Dawn

The preceding chapter pointed out that, ever since the 1980s, NASA has been in dire need of a major objective that only it—and it alone—could perform. The short-lived Space Exploration Initiative to travel to Mars (among other things) was clearly intended as one such objective. Even though that program died a quiet death (if it could truly be said to have had a life at all) in 1990,[38] many space enthusiasts continue to press for Mars as NASA's next big destination.[39]

Somewhat like the rocket societies of the 1920s and 30s (see chapter 3), organizations like the Mars Society contains a large number of members with scientific and technical backgrounds (in fact, the Mars Society president, Robert Zubrin, is a trained engineer). Thus, in addition to the usual lobbying and campaigns to generate "grassroots" support, some Mars proponents have also been working to develop—and more important, lower the expense of—the relevant technologies. Believing (correctly) that a major reason Congress refused to approve funding for SEI was the multi-hundred-billion dollar price tag (estimated in some quarters to be as high as $450 billion[40]), supporters of a mission to the red planet (or, as some of their literature would have it, the "New World") have mounted a major effort to bring the costs down to a "politically acceptable" level.

One such scenario, put forward by Zubrin himself, is called "Mars Direct." To date, space missions involving humans always carry along all of their fuel, including that needed for the trip home. This represents a massive weight requirement, which in turn either limits the size of the spacecraft or drives up the design costs. Under the Mars Direct proposal, propellant for the return flight would be manufactured on Mars, from chemicals already known to exist in abundance there. A spacecraft would therefore only have to carry half as much fuel as most current plans, a savings in both weight and cost. Using these and other innovative—but not necessarily exotic—technologies, Zubrin estimates that a Martian mission could be undertaken within the next decade for a total cost $20 billion.[41]

Unfortunately, the analysis presented over the past eight chapters suggests that such efforts, well meaning though they may be, are not likely to be sufficient. It may well be correct that the funding requirements of Mars Direct are—adjusted for inflation—lower than those for Apollo in the 1960s. It must be remembered, however, the elected officials in 1961 were willing to support the moon landing despite the cost, because they had defined it as an issue of immense national significance. To paraphrase Vice President Johnson at the time, it is difficult to worry about cost if one believes that national survival is at stake. Certainly many of those same officials just a few years later balked at the cost of the proposed follow-ups to Apollo. Still, their behavior up to that point strongly suggests that they would have appropriated the necessary funds had they continued to view the program as any sort of national priority.

Zubrin's proposal, even assuming that the technology worked as promised (something that, as NASA has discovered, is never a given), addresses only one side of the political equation. Even the very cheapest Mars program will still be among one of the government's larger budget items. It is therefore only likely to receive approval if members of Congress, as well as executive branch officials, view it as being worth the cost. In short, what sort of issue definition would persuade policymakers to approve an average appropriation of $2 billion, and continue to do so for more than a decade?

Finally, the most recent developments suggest that the premise of this section—that is, that a human flight to Mars is something that only NASA could accomplish—may not be correct. Consistent with the trend that began in the 1980s (see chapters 7 and 8), there is increasing discussion of bypassing the agency altogether and relying on the private sector. Zubrin himself has reportedly proposed that Congress offer a "Mars prize" of $20 billion to the business consortium that reaches the planet (and presumably returns safely) first.[42] There is little reason to suppose that Congress will act on this idea—or that private industry will be gearing up for a run at Mars—anytime soon. Still, the fact that such a proposal is being taken at all seriously suggests that NASA may no longer have a monopoly on even Apollo-style missions.

Helpful Leadership

James Webb was almost certainly correct when he told President Johnson that it was up to the government to tell NASA what to do, not the other way around.[43] Part of Webb's thinking—also reflected in some later political writing[44]— is that the significant resource requirements associated with spaceflight make it essential that there be a clear consensus among policymakers before any ambitious space plans can get underway. Kennedy's "moon speech" seems, at first glance, to represent such an approach. That address, in fact, has become not just a model for how new space initiatives ought to be unveiled, but in some quarters it is seen as the very definition of "leadership" in space.

Unfortunately, the subsequent success of Apollo (at least to the extent of getting funding and actually reaching the moon) has resulted in the Kennedy speech looking, in retrospect, far more significant than it actually was. Of course, it *was* significant; without it there probably would not have been anything like a Project Apollo for many years.[45] One must be careful, however, not to confuse a *necessary* condition with a *sufficient* one. The president's 1961 pronouncement was absolutely necessary to getting the program underway. It was not, in and of itself, sufficient: as chapters 4 and 5 point out, there were, in the late 1950s and early 1960s, a number of political, economic, and social factors contributing to the space program's heightened political saliency. Without those (or similar) conditions present, no presidential declaration, no matter how lofty or poetic, is enough to guarantee a multibillion dollar, multiyear expenditure. This can clearly be seen in some of the subsequent efforts to emulate Kennedy: for years, Reagan's space station proposal was in great danger of cancellation, and, as already noted, Bush's speech on behalf of SEI caused barely a ripple.

This does not mean that there is no role for presidential leadership. Perhaps the most significant contribution a future chief executive can make to NASA is provide an explicitly executive function. Many of the policy prescriptions addressed to the agency are couched as negatives, that is, as things that NASA should *not* do. It should not, for example, try to compete with the private sector, just as it should not be involved with space operations. Even some ostensibly prescriptive statements are really just backhanded negatives, as in the view that NASA should "carry out only those activities that cannot be accomplished by commercial means," or else too vague to be of any use, such as stating that "there is some role for NASA" without specifying what that might be.

What would surely be more helpful would be some type of actual policy guidance, a set of specific charges for the agency. Policymakers should, in conjunction with leaders from commercial space industry and other stakeholders, establish each group's jurisdictions and areas of responsibility.[46] This would, among other things, force both elected officials and the NASA leadership to return—literally for the first time since the Space Act was written—to the basic question of what NASA is, or rather, what it is for. Although not as flashy or headline grabbing as a grand, public speech, it is the sort of administrative reform that might, in the long run, be a far better service to the agency and to the program.

Back to the Future

At least one author has described the Space Exploration Initiative as a failed attempt to revive the 1960s, "Apollo-like" approach to space policy:[47] a grand, highly challenging long-term objective (or, since SEI called for a permanent lunar base and a human mission to Mars, *two* grand objectives) requiring the development of a great many new technologies, a bold declaration of these goals by the president in a high-profile address, and, of course, hefty increases in the NASA budget (although the evidence suggests that President Bush was not made fully aware of how hefty an increase would be needed[48]). As chapter 8 pointed out, however, aside from a few "true believers" in the Bush administration, NASA, and elsewhere, very few people—in or out of government—were in a 1960s frame of mind.

Nevertheless, the comparison is somewhat ironic, since, in one very important respect, space-related activity in the United States around this time really did resemble that of an earlier period. Due in large part to the policies of his administration, by the end of President Reagan's second term the United States could no longer be said to have a space program—like the late 1950s, it had a number of space programs, the largest of which (at least in budgetary terms) was run by the military. Of course, there are also significant differences between the two periods, some of which have had rather serious implications for NASA. By all accounts, the struggle between the army, navy, and air force for control over American space policy 40 years earlier was at times quite intense. Indeed, it often looked like they were competing harder against each other than against the Soviet Union (a 1957 Herblock cartoon depicts an officer watching *Sputnik* and saying to a colleague with relieved expressions, "Whew! At first I thought it was sent up by one of the other services"[49]). Sometimes, as chapters 3 and 4 noted, this arrangement led to unnecessary duplication of facilities and other inefficiencies. Nevertheless, it is quite clear that, despite their differences, all participants agreed on the program's basic goals.

It should come as no surprise that a half century of steady technological development, in conjunction with the emergence of a large and diverse private sector, has led to a rapid increase in both the number of organizations and individuals involved in space-based activities and the number—and type—of goals and objectives they are pursuing.[50] Moreover, it can be shown quite readily that, as the number and diversity of goals increases, the more likely they are to come into conflict. Even organizations whose objectives are ostensibly identical may in fact have nothing in common whatsoever: two firms could be "seeking to make a profit," but a large NASA contractor—say, one involved in major International Space Station construction—will behave very differently from a provider of commercial launch services (which, at least until the mid-1980s, might actually have been in competition with NASA).[51]

In short, the closing decade of the 20th century saw the emergence of a division within the space community,[52] one that would have been unimaginable a generation ago. For most of its history, NASA was the space program: for policymaker and citizen alike, being "pro-space" meant—automatically—being "pro-NASA." During the 1960s, those few individuals who tried to sound supportive of spaceflight "in general" while criticizing a specific NASA project (see, for example, Senator Fulbright's arguments against Apollo in chapter 4) tended to come off as somewhat disingenuous, if not completely eccentric. A decade later, when the program's emphasis switched from large-scale spectacle to "applied" missions that would "improve life here on earth," there was never any serious question, let alone any debate, about who would direct those missions.[53]

From the mid-1980s onward, however, NASA was increasingly likely to find itself lined up against private companies, influential members of Congress, and various other federal officials and agencies. And, in these contests, it was becoming more common for the agency to find itself cast as a villain.

By the end of the decade, however, the agency's virtual monopoly status had been largely eroded. As the last chapter pointed out, the dominant view of space community in the final decade of the 20th century was that space policy was not necessarily the same as NASA policy. In fact, a growing number of private entrepreneurs, conservative scholars and commentators, and public officials began to see the agency as an impediment to their vision of what constitutes sound space policy.

Put another way, largely because of major policy changes in the 1980s, the number of participants in space-related activities has grown sharply. Just what NASA's role should be in this new and expanding constellation of space policy actors is not at all clear. It has been many years since NASA had the only ticket into space, but there is still less agreement than ever before as to what destination that particular ticket should have stamped on it.

Perhaps the largest difference between the program(s) now and that of 50 years ago is in the nature of space activity itself. There is a great deal of irony that, in the midst of all of this debate and conflict, spaceflight itself is more robust than at any other time in history. By the century's end, humanity was involved in space (albeit usually not by going there) and using space-related products as never before. Moreover, just as NASA had promised a decade earlier, space operations themselves really appear to be on the verge of becoming routine (or as close to routine as such things as space launches can be). It seems obvious that the human race is in space to stay. What is far less clear is what is to become of the organization that made that development possible.

Epilogue: 2001–2003

One of the more vexing problems facing authors who write about science- and technology-based issues—and the one most frustrating for their publishers—is deciding on closure. Spaceflight, cloning, telecommunications, computers, and other such fields undergo continuous development, with regard to the policies that surround them, as well as in the technologies themselves. As a result, any analysis of these programs, especially in book form, run the risk of becoming seriously outdated even before appearing in print. There is no really satisfactory solution to this problem other than simply declaring "the end" at some rational-sounding, but, in reality, completely arbitrary date.[54] It is in this spirit that this book has set as its purpose an examination of the historical debate over NASA's mission from its founding in 1958 (along with a bit of background from the 1930s and 1940s) through the end of 2000.

The first few years of the 21st century, however, have been marked by a number of incidents that are just too important to ignore. Each has had a profound impact on world space policy in general, and NASA's programs in particular, and will almost certainly continue to influence our activities in space for some time to come. This chapter will therefore conclude with a brief look at some of these events.

Taking the Bad with the Good

The first—at least chronologically—was an economic slowdown that began in early 2001 (some economists date it a bit earlier), and which turned into a full-fledged recession the following year. By the end of 2002, a large number of space-related companies (most notably those pursuing new launch technologies) had either declared bankruptcy or drastically scaled back their operations.[55] Satellite operators, who during the 1990s had been anticipating continued growth in the wake of an expected booming telecommunications market, suffered through "a miserable year . . . perhaps the worst since the satellite sector came to be known as an industry."[56] The lower demand for satellite technology led, in turn, to a sharp decline in the demand for commercial launches.[57] By mid-2003, a joke began circulating within the commercial space community that "the best way to make a small fortune in space is to start with a large fortune."[58]

Those firms that managed to survive faced many of the serious problems that commonly occur in a depressed market.[59] First, the cost-cutting measures imposed by some companies seeking to lower their operating expenses has led, in some cases, to a decline in product quality and even, in a few instances, to costly failures.[60] Such incidents, of course, often lead to further market erosion. In addition, the industry began to experience an exodus of personnel. Throughout the period 2001–2003, the perceived lack of opportunity led a significant number of older aerospace engineers to retire, and sent the younger ones in search of more lucrative industries (unfortunately, since the recession affected virtually all areas of high technology, it is not at all clear what—or where—those might be).[61] Finally, despite all the confident talk of the emerging private space sector that dominated the 1990s, by 2003 almost every commercial space enterprise still in operation was heavily dependent on government—particularly the military (see below)—as its primary customer.[62]

Viewed in retrospect, none of this should have been terribly surprising. According to free market advocates, the reason that private firms are supposed to operate so much more efficiently than government organizations is because the latter have no fear of competition or bankruptcy. What this also means, of course, is that, unlike NASA, space-based businesses *can* go bankrupt. In fact, the exceedingly high costs—and high risk—associated with this industry seem to render it especially sensitive to forces in the larger economy.[63] Put another way, just as markets rise, as they did during the 1990s, they can also fall. Moving the bulk of space activity out of the public sector, as many were calling for in the 1990s, may (arguably) have made it more economically efficient, increased the number of participants, and spurred innovation, but it also had the effect of tying the future of spaceflight to the vagaries of the business cycle. Public programs, for all their limitations, are (at least relatively) recession proof.

This should not be interpreted as a call for returning to a 1960s-style program, or for reimposing a government monopoly on space. The point, rather, is simply that the great shift in U.S. space policy that began in the 1980s, whatever its benefits, also contained a certain measure of risk, a fact that was not fully appreciated at the time, or during the decade that followed. Moreover, the economy, and commercial space, will almost certainly rebound eventually.

It is important to note, however, that while the decline in the private space industry has greatly increased its reliance on—and thus the importance of—public programs, this does not appear to have benefited NASA in any meaningful way. As noted above, practically all of the increased government spending in this area has come from the armed services. The reason for this is to be found in the second major event of our young century.

A New Definition: Homeland Security

Chapter 2 noted that policymakers' perceptions of public issues and their relative importance can be altered—sometimes drastically—by some unexpected incident. Few such events, however, have had as profound an impact, across as many policy areas, as the terrorist attacks of September 11, 2001. It is worth noting, in passing, the many similarities between that tragedy and the panic that surrounded *Sputnik* nearly 45 years earlier. Of course, the differences are quite substantial as well. In particular, 9/11 was a direct attack that killed more than 3,000 people. *Sputnik* never posed any threat in and of itself. As chapter 4 pointed out, the danger—such as it was—came from what many feared the Soviet satellite represented. Moreover, unlike the "satellite crisis," which proved politically damaging to President Eisenhower, the attacks on New York and Washington caused many Americans to rally behind President Bush.

Even so, there are numerous parallels.[64] In both cases, the United States faced ideologically driven opponents seeking to demonstrate (primarily to their own adherents) that their belief system was superior to that of a supposedly "declining" or "decadent" superpower. As in 1957, most Americans were left with feelings of vulnerability that most had never known before. Some Republican politicians tried to assign blame for both events to "neglect" on the part of the preceding Democratic administrations, while Democrats (including a number running for president), criticized Republican officials (although—at least not at first—the president directly) for "not doing enough" about the crisis. Within a few years of each event, the incumbent president took a historic step—Project Apollo for John F. Kennedy; the preemptive war in Iraq for George H. W. Bush—that was (at least initially) approved by a majority of the public, but which critics saw as expensive, risky, and largely unrelated to the original problem.

Finally, and particularly relevant to the present discussion, 9/11, like *Sputnik*, triggered a massive retaliatory effort involving a significant government reorganization and a wholesale change in public priorities (not unlike that described in chapter 4). Chief among these has been a heightened emphasis on homeland security. Included in what can only be termed yet another redefinition of government policy has been the nation's R&D agencies.[65]

Obviously, space technology has made a major contribution to this effort. Surveillance satellites have attempted to locate and track terrorist groups, while intelligence agencies have made use of space-based systems to try to listen in on cell phone and wireless communications.[66] Satellites also played a vital role in the invasions of Iraq and Afghanistan (in the latter case, where the mountainous terrain blocks conventional communications networks, satellite-based systems were essential in keeping military units in contact with one another).[67] Given these facts, it is hardly surprising that the Bush administration's spending on space has heavily emphasized defense applications: by FY 2003, the budget for military space had grown to more than $18 billion.[68]

These new government priorities, in combination with the return of very large federal deficits and the *Columbia* accident (discussed below), has made it even more difficult for NASA to secure funding for new projects.[69] A few space proponents have sought to incorporate large civil space programs into the new definition—in mid-2003, an editorial writer for the *Wall Street Journal* tried to identify a human mission to Mars as "relevant to the battle of civilizations that we find ourselves in now," claiming that "such an adventure would be an emblem of the forward-looking, empirical, and ambitious Western outlook"[70]—but, in the present political environment, it is difficult to imagine such efforts succeeding. Indeed, it is unclear how the agency could even begin to redefine more than a small portion of its programs—and clearly nothing as massive as the space station—to conform to these new government priorities. In short, the events of September 11 and beyond have pushed NASA even further into irrelevancy.

Columbia

On Saturday morning, February 1, 2003, during the reentry of what until that moment had been a routine mission, *Columbia*, America's—and the world's—first reusable space shuttle, broke apart in the skies over Texas, killing all seven crew members. It was the shuttle program's second catastrophic failure in 113 flights. Most of the evidence seemed to point to a large piece of foam insulation that struck the wing of the orbiter during liftoff, cracking the thermal tiles that protect a shuttle during its descent through the atmosphere, as the primary cause of the accident.[71] As it did after the loss of *Challenger*, NASA instituted a number of managerial and procedural changes—such as reinstituting the practice (abandoned early in the program's history) of visually surveying the condition of the orbiter, using onboard cameras and military surveillance satellites, before attempting reentry—aimed at preventing a recurrence.[72] The agency publicly predicted that the remaining vehicles in the fleet would begin operating again sometime in 2004.[73]

Unfortunately, the problems raised by this disaster go far beyond the issue of its immediate causes and remedies. Although the loss of a spacecraft with all hands would of course be a terrible tragedy in any event, it is difficult to imagine a worse time for such a thing to happen. To begin with, as already noted, government's priorities on space have shifted heavily toward defense, an area from which NASA is legally excluded. The agency was already faced with seriously declining resources,[74] at a time when the federal government is again confronting significant budget deficits. Finally, *Columbia* represented one-quarter of NASA's shuttle fleet. After the loss of *Challenger* in 1986, the Reagan administration approved the construction of a replacement orbiter. There is virtually no chance that President Bush will approve such a venture.[75] NASA has explored a number of options for a new system for launching humans into space, but this effort is unlikely to produce any tangible results anytime in the near future.[76] Meanwhile, the accident has had a serious impact on the construction and operation of the International Space Station. Until the shuttles return to flight, ISS personnel will have to depend on the Russian *Soyuz* spacecraft and robot vehicles for supplies and crew rotation.[77]

For NASA, however, the most troublesome issue to follow in the aftermath of *Columbia* may be a renewed, if not reinvigorated, debate over the agency's overall purpose. As happened after *Challenger*, critics are questioning the cost and risk involved in sending humans into space, a debate that has focused on the International Space Station as well as the shuttle program.[78] Although NASA, with the apparent support of the Bush administration, is committed to returning the STS to flight in 2004, the future of human spaceflight over the long term seems anything but certain. And, if yet another tragic accident were to occur. . . .

What Is the Right Path?

The last three chapters have shown that the dominant trend in U.S. space policy during the 1980s and 1990s has been a growing divergence between NASA and its programs and space activity in general. Put another way, there was during those two decades a rapid, if not explosive, growth in the human use of space, by both public and private actors, in which the agency played, at best, a supporting role. Almost everything that has happened in the opening years of the 21st century has, if anything, accelerated this trend. Space technology is indeed a critical element in one of the federal government's highest priorities, the war on terror, but is almost exclusively a DOD responsibility. Not only are NASA's largest projects facing renewed scrutiny and criticism, but the loss of *Columbia*, combined with its other resource problems—and in a time of severe budget constraints—has left the organization with a greatly diminished capacity for pursuing its programs. Barring yet another major shift in the prevailing definition of space policy (which at the present time appears highly unlikely), the U.S. government's endeavors in civilian space exploration seem virtually certain to remain static for some time to come, and may even shrink further.

This book began with the observation that the United States has made a substantial investment in space. NASA, for all of its problems, still maintains a large, highly trained (albeit shrinking and aging) workforce, numerous field centers, and an extensive infrastructure devoted to space research and development. Although shrinking, its annual budget still runs into billions of dollars. Yet the answer to the question posed in chapter 1 seems more elusive than ever: What is NASA for? It is well past time for policymakers, stakeholders, and anyone else with a serious interest in our future in space to begin coming up with an answer.

Notes

Chapter 1

1. U.S. National Aeronautics and Space Administration, *Aeronautics and Space Report of the President, Fiscal Year 1998 Activities* (Washington, DC: NASA, 1999), app. E-1B. Figures are adjusted for inflation.

2. See, for example, Gregg Easterbrook, "Lost in Space," *Washington Monthly* (April 1987): 48–54, and "The Spruce Goose of Outer Space," *Washington Monthly* (April 1980): 32–48. In addition, the widely held belief that spaceflight was responsible for the creation of Teflon, Velcro, and Tang turns out to be only a myth.

3. The controversy regarding spin-offs is not over whether they really exist, but whether they are a sufficient reason, in and of themselves, to spend billions of dollars on space projects. More recently, this issue has spread to the discussion of U.S. research and development policy generally. See John A. Alic, Lewis M. Branscomb, Harvey Brooks, and Ashton Carter, *Beyond Spinoff: Military and Commercial Technologies in a Changing World* (Boston: Harvard Business School Press), 1992; and David H. Guston, *Between Science and Politics: Assuring the Integrity and Productivity of Research* (New York: Cambridge University Press), 2000, particularly chap. 5.

4. More recently, the debate has also turned to who should be responsible for doing it (see chapters 7 and 8).

5. The Departments of Defense (DOD), Commerce (DOC), and Energy (DOE), among others, all have space projects of one kind or another. Moreover, there have been periods over the past 50 years when DOD was actually the nation's largest space agency, at least as measured by funding. In 1988, for example, DOD's space budget was almost double that of NASA's. See U.S. National Aeronautics and Space Administration, *Aeronautics and Space Report*.

6. Although not as visible as human spaceflight, NASA's (relatively) smaller automated space science programs—including the Hubble Space Telescope, the *Galileo* (Jupiter) and *Cassini* (Saturn) probes, and the Halley's comet intercept mission (which was eventually canceled)—have also come in for their share of criticism.

7. It is always possible, of course, that some new technical innovation (or, more likely, a series of innovations) could change this by lowering the costs and reducing the risks.

8. National Aeronautics and Space Act, Public Law 85-568, sec. 102 (c), 1–8. The full text of the act can be found in John M. Logsdon, ed., *Exploring the Unknown: Selected Documents in the History of the U.S. Civil Service Program*, vol. 1, *Organizing for Exploration* (Washington, DC: NASA, 1995), 334–345.

9. Ibid., sec. 203 (b) (5).

10. Ibid., sec. 203 (a) (3).

11. This omission was addressed in a 1984 amendment. See the discussion in chapter 7.

12. U.S. National Aeronautics and Space Administration, *NASA Strategic Plan, 2000* (Washington, DC: NASA, 2000), 1.

13. An attempt to describe some of this complexity can be found in W. D. Kay, *Can Democracies Fly in Space? The Challenge of Revitalizing the U.S. Space Program* (Westport, CT: Praeger, 1995).

14. This includes the organizational development of NASA's predecessor organizations, particularly the National Advisory Council on Aeronautics. See Howard E. McCurdy, *Inside NASA: High Technology and Organizational Change in the U.S. Space Program*, New Series in NASA History (Baltimore: Johns Hopkins University Press), 1993.

15. See Alexander J. Morin, *Science and Politics* (Englewood Cliffs, NJ: Prentice-Hall, 1993), chap. 5.

16. Whether such research is actually worth the current cost of getting into orbit is a matter of some dispute.

17. There are even some who envision a time when all factories could be moved off the earth's surface, thereby eliminating industrial pollution forever. See, for example, G. Harry Stine, *The Space Enterprise* (New York: Ace Books), 1980.

18. See Arthur C. Clarke, *The Promise of Space* (New York: Harper and Row, 1968). A similar claim is often made by proponents of the Search for Extraterrestrial Intelligence program, who often speak of the "new attitude" they believe will follow from the discovery that "we are not alone."

19. Bryan Bunch with Alexander Hellmans, *The History of Science and Technology: A Browser's Guide to the Great Discoveries, Inventions, and the People Who Made Them From the Dawn of Time to Today* (Boston: Houghton-Mifflin, 2004), 719.

20. Paul Kallender, "Officials Call for Japan to Establish Manned Space Program," *Space News*, 17 February 2003, 17.

21. A task which must, of course, also take into account who is receiving those benefits.

22. See, for example, Roger D. Launius and Howard E. McCurdy, introduction to *Spaceflight and the Myth of Presidential Leadership* ed. Roger D. Launius and Howard E. McCurdy, 1–14 (Urbana: University of Illinois, 1997).

23. See, among many others, Rip Bulkeley, *The Sputniks Crisis and Early United States Space Policy* (Bloomington: Indiana University Press, 1990); Edwin

Diamond, "Sputnik," *American Heritage* 48.6 (1997): 84–93; Robert A. Divine, *The Sputnik Challenge: Eisenhower's Response to the Soviet Satellite* (New York: Oxford University Press, 1993); Roger D. Launius, "Eisenhower, Sputnik, and the Creation of NASA: Technological Elites and the Public Policy Agenda," *Prologue* 28.2 (1996): 127–143; John M. Logsdon, *The Decision to Go to the Moon: Project Apollo and the National Interest* (Chicago: University of Chicago Press, 1970); Walter A. McDougall, . . . *the Heavens and the Earth: A Political History of the Space Age* (New York: Basic Books, 1985); John M. Logsdon, "The Space Shuttle Program: A Policy Failure?" *Science*, 30 May 1986, 1099–1105; and Howard E. McCurdy, *The Space Station Decision: Incremental Politics and Technological Choice*, Johns Hopkins New Series on NASA History (Baltimore: Johns Hopkins University Press, 1990).

24. Actually, as the chapter will show, the agency's problems actually began a few years before the first lunar mission.

Chapter 2

1. David Dery, *Problem Definition in Policy Analysis* (Lawrence: University Press of Kansas, 1984), 16–17.

2. Frank R. Baumgartner and Bryan D. Jones, "Attention, Boundary Effects, and Large-Scale Policy Change in Air Transportation Policy" in *The Politics of Problem Definition: Setting the Policy Agenda*, ed. David A. Rochefort and Roger W. Cobb, 50–66 (Lawrence: University of Kansas Press, 1994).

3. John Portz, "Plant Closings, Community Definitions, and the Local Responsem" in *The Politics of Problem Definition: Setting the Policy Agenda*, ed. David A. Rochefort and Roger W. Cobb, 32–49. (Lawrence: University of Kansas Press, 1994).

4. David Howard Davis, *Energy Politics*, 4th ed. (New York: St. Martin's, 1993), chap. 2.

5. Caroline J. Acker, "Stigma or Legitimation? A Historical Examination of the Social Potentials of Addiction Disease Models," *Journal of Psychoactive Drugs* 25 (1993): 193–205; and Peter Reuter, "Hawks Ascendant: The Punitive Trend of American Drug Policy," *Daedalus* 121 (1995): 15–52.

6. James M. Griffin and Henry B. Steele, *Energy Economics and Policy*, 2nd ed. (Orlando, FL: Academic, 1986), 168.

7. David A. Rochefort and Roger W. Cobb, "Framing and Claiming the Homelessness Problem," *New England Journal of Public Policy* 8 (1992): 49–65.

8. David A. Rochefort and Roger W. Cobb, "Instrumental versus Expressive Definitions of AIDS Policymaking," in *The Politics of Problem Definition: Setting the Policy Agenda*, ed. David A. Rochefort and Roger W. Cobb, 159–181 (Lawrence: University of Kansas Press, 1994). See also Sandra Panem, *The AIDS Bureaucracy*. (Cambridge, MA: Harvard University Press, 1988); John Street, "Policy Advice in an Established Advice Structure: AIDS Advice through the British Department of Health," in *Advising West European Governments: Inquiries, Expertise, and Public Policy*, ed. B. Guy Peters and Anthony Barker, 151–164 (Edinburgh: Edinburgh University Press, 1993).

9. Beryl A. Radin and Willis D. Hawley, *The Politics of Federal Reorganization: Creating the U.S. Department of Education* (New York: Pergamon, 1988).

10. Davis, *Energy Politics*, chap. 6. See also Brian Balogh, *Chain Reaction: Expert Debate and Public Participation in American Commercial Nuclear Power, 1945–1975* (New York: Cambridge University Press, 1991); and David E. Lilienthal, *The Journals of David E. Lilienthal*, vol. 2, *The Atomic Energy Years, 1945–1950* (New York: Harper and Row, 1964).

11. See also Anthony Barker and B. Guy Peters, "Science and Government," in *The Politics of Expert Advice: Creating, Using, and Manipulating Scientific Knowledge for Public Policy*, ed. Anthony Barker and B. Guy Peters, 1–16 (Pittsburgh: University of Pittsburgh Press, 1993); Cheol H. Oh, "Explaining the Impact of Information on Problem Definition: An Integrated Model," *Policy Studies Review* 15.4 (1988): 109–136.

12. David A. Rochefort and Roger W. Cobb, "Problem Definition: An Emerging Perspective" in *The Politics of Problem Definition: Setting the Policy Agenda*, eds. David A. Rochefort and Roger W. Cobb, ___, especially 26 (Lawrence: University of Kansas Press, 1994). See also Andrea L. Bonnicksen, *Crafting a Cloning Policy: From Dolly to Stem Cells* (Washington, DC: Georgetown University Press, 2002); S. Begley, "Cellular Divide," *Newsweek*, 9 July 2001, 24; and L. Goodstein, "Abortion Foes Split over Plan on Stem Cells," *New York Times*, 12 August 2001, A1, A22.

13. Goodstein, "Abortion Foes Split"; Laurie McGinley and Anne Fawcett, "Patients and Abortion Foes Clash on Stem-Cell Research," *Wall Street Journal*, 21 June 1999, A28; and Bob Davis, "GOP Avoids Abortion for Now, but Science Is Stirring the Debate," *Wall Street Journal*, 1 August 2000, A1.

14. Rochefort and Cobb, "Problem Definition."

15. Ellen Frankel Paul, "Sexual Harassment: A Defining Moment and Its Repercussions," in *The Politics of Problem Definition: Setting the Policy Agenda*, ed. David A. Rochefort and Roger W. Cobb, 67–97 (Lawrence: University of Kansas Press, 1994).

16. See Stanley Karnow, *Vietnam: A History* (New York: Vintage, 1983); Peter Braestrup, *Big Story: How the American Press and Television Reported and Interpreted the Crisis of Tet 1968 in Vietnam and Washington* (New Haven, CT: Yale University Press, 1983); Robert R. Tomes, *Apocalypse Then: American Intellectuals and the Vietnam War, 1954–1975* (New York: New York University Press, 1998); and Anthony O. Edwards, *The War in Vietnam* (Westport, CT: Greenwood, 1998).

17. Balogh, *Chain Reaction*.

18. Christopher J. Bosso, *Pesticides and Politics: The Life Cycle of a Public Issue* (Pittsburgh: University of Pittsburgh Press, 1987).

19. William H. Riker, *The Art of Political Manipulation* (New Haven, CT: Yale University Press, 1986). See also his "Implications from the Disequilibrium of Majority Rule for the Study of Institutions" *American Political Science Review* 74.2 (1980): 432–446.

20. Amal Kawar, "Issue Definition, Democratic Participation, and Genetic Engineering," *Policy Studies Journal* 17.4 (1989): 719–744. See also Patrick W. Ham-

lett, *Understanding Technological Politics: A Decision-Making Approach* (Englewood Cliffs, NJ: Prentice-Hall, 1991), chap. 10; and Sheldon Krimsky and Roger Wrubel, *Agricultural Biotechnology and the Environment: Science, Policy, and Social Issues* (Urbana: University of Chicago Press, 1996).

21. L. Christopher Plein, "Popularizing Biotechnology: The Influence of Issue Definition," *Science, Technology, and Human Values* 16.4 (1991): 474–490.

22. One of the first attempts by a political scientist to search for a pattern to these changes was John W. Kingdon, *Agendas, Alternatives, and Public Policies* (Boston: Little, Brown, 1984). See also the discussion in the next section. It does seem likely that terrorism will be a high priority for quite some time, although one could argue that this is simply a subset of national defense.

23. Davis, *Energy Politics*.

24. Fred Pelka, *The ABC-CLIO Companion to the Disability Rights Movement* (Santa Barbara, CA: ABC-CLIO, 1997).

25. John Portz, "Problem Definitions and Policy Agendas: Shaping the Educational Agenda in Boston," *Policy Studies Journal* 24.3 (1996): 371–386.

26. Rochefort and Cobb, "Problem Definition," 21–22.

27. See Linda R. Cohen and Roger G. Noll, "Synthetic Fuels from Coal" in *The Technology Pork Barrel*, ed. Linda R. Cohen and Roger G. Noll, 259–321 (Washington, DC: Brookings Institution, 1991). It should be noted that the program had also encountered significant technical and economic problems.

28. For a brief general history of the SSC program, see David Ritson, "Demise of the Texas Supercollider," *Nature*, 16 December 1993, 607–610. For descriptions of its financial and administrative troubles, see Malcolm W. Browne, "Supercollider's Rising Cost Provokes Opposition." *New York Times*, 29 May 1990, C1, C6; Christopher Anderson, "University Consortium Faulted on Management, Accounting," *Science*, 9 July 1993, 157 and "DOE Pulls Plug on SSC Contractor," *Science*, 12 August 1993, 822–823. For DOE's and the physics community's views on the project, see U.S. Department of Energy, Office of Energy Research, Division of High Energy Physics, *High Energy Physics Advisory Panel's Subpanel on Vision for the Future of High Energy Physics* (Washington, DC: DOE, 1994).

29. Sheldon L. Glashow and Leon M. Lederman, "The SSC: A Machine for the Nineties," *Physics Today* (March 1985): 28–37; James W. Cronin, "The Case for the Super Collider," *Bulletin of the Atomic Scientists* (May 1986): 8–11; Colin Norman, ". . . But They Endorse the Supercollider," *Science*, 9 May 1986, 705; and Mark Crawford, "Reagan Okays the Supercollider," *Science*, 6 February 1987, 625.

30. See Bryan D. Jones, *Reconceiving Decision-Making in Democratic Politics* (Chicago: University of Chicago Press, 1994), especially chap. 4. For a slightly different account of the SSC's redefinition (from a domestic to an international program), see W. D. Kay, "Particle Physics and the Superconducting Super Collider: The Collaboration That Wasn't," in *International Collaboration in Science and Technology among Advanced Industrial Societies*, ed. Vicki L. Golich and W. D. Kay (London: Routledge, forthcoming).

31. Bosso, *Pesticides and Politics*.

32. W. D. Kay, "Congressional Decision-Making and Long-Term Technological Development: The Case of Nuclear Fusion," in *Science, Technology, and Politics: Policy Analysis in Congress*, ed. Gary Bryner, 87–106 (Boulder, CO: Westview, 1992).

33. Griffin and Steele, *Energy Economics and Policy*, 168.

34. U.S. Congress, Office of Technology Assessment, *Nuclear Power in an Age of Uncertainty* (Washington, DC: OTA, 1984), chap. 8.

35. Plein, "Popularizing Biotechnology."

36. See, for example, Graham T. Allison, *The Illusion of Choice: Explaining the Cuban Missile Crisis* (Boston: Little, Brown, 1971), 79–81, 97; and John D. Steinbruner, *The Cybernetic Theory of Decision: New Dimensions of Political Analysis* (Princeton, NJ: Princeton University Press, 1974).

37. W. D. Kay, "John F. Kennedy and the Two Faces of the U.S. Space Program, 1961–1963," *Presidential Studies Quarterly* 28.3 (1998): 573–586, and "Problem Definition and Policy Contradiction: John F. Kennedy and the 'Space Race,'" *Policy Studies Journal* 31.1 (2003): 53–69.

38. Balogh, *Chain Reaction*.

39. See Charles S. Bullock III, James E. Anderson, and David W. Brady, *Public Policy in the Eighties* (Monterrey, CA: Brooks-Cole, 1983), chap. 5.

40. William P. Browne, *Cultivating Congress: Constituents, Issues, and Interests in Agricultural Policymaking* (Lawrence: University of Kansas Press, 1995). See also Louis Ferleger and William Lazonick, "The Managerial Revolution and the Developmental State: The Case of U.S. Agriculture," *Business and Economic History* 22.2 (1993): 67–98.

41. Discussions of the latter can be found in U.S. Congress, Office of Technology Assessment, *Plants: The Potentials for Extracting Protein, Medicines, and Other Useful Chemicals—Workshop Proceedings* (Washington, DC: OTA, 1983).

42. Charles T. Goodsell, *The Case for Bureaucracy: A Public Administration Polemic*, 3rd ed. (Chatham, NJ: Chatham House, 1985), chap. 3.

43. See Henry Mintzberg, *The Structure of Organizations* (Englewood Cliffs, NJ: Prentice-Hall, 1979), and *Structure in Fives: Designing Effective Organizations* (Englewood Cliffs, NJ: Prentice-Hall, 1981).

44. J. D. Thompson, *Organizations in Action* (New York: McGraw-Hill, 1967).

45. Jerry L. Mashaw, *Bureaucratic Justice: Managing Social Security Disability Claims* (New Haven, CT: Yale University Press, 1983).

46. U.S. Congress, General Accounting Office, *The Department of Energy: A Framework for Restructuring DOE and Its Missions* (Washington, DC: GAO, 1995). Of course, DOE is hardly alone in this. See chapter 9.

47. U.S. Department of Energy, *Department of Energy, 1977–1994: A Summary History*, Energy History Series (Washington, DC: DOE, 1994), app. 5-A.

48. Ibid., 74. See also U.S. Congress, General Accounting Office, *Environment, Safety, and Health: Status of DOE's Reorganization of its Safety Oversight Function* (Washington, DC: GAO, 1990).

49. The fiscal year (FY) 1995 budget for Defense programs was less than 30 percent of DOE's total, the lowest level in the department's history. See U.S. Department of Energy, *Department of Energy, 1977–1994*.

50. See "Summit Focuses on Vision to Guide Rocky Flats Cleanup," *DOE This Month* 18.4 (1995): 14; and "Cleanup Accelerated at Hanford Defense Plants," *DOE This Month* 18.2 (1995): 5. See also Dianne Rahm, "Coming into Environmental Compliance: DOE and the Nuclear Weapons Complex" (paper presented at the annual meeting of the American Political Science Association, Atlanta, GA, 2–5 September 1999).

51. U.S. Congress, General Accounting Office, *The Department of Energy*; and Rahm, "Coming into Environmental Compliance."

52. Charles Perrow, "The Analysis of Goals in Complex Organizations," *American Sociological Review* 26 (1961): 854–866. See also Lawrence B. Mohr, "The Concept of Organizational Goal," *American Political Science Review* 67 (1973): 470–481 (where the terms are "official" and "reflexive" goals).

53. Anthony Downs, *Inside Bureaucracy* (Boston: Little, Brown, 1967). For a discussion of this idea as it relates to NASA, see Howard E. McCurdy, *Inside NASA: High Technology and Organizational Change in the U.S. Space Program* (Baltimore: Johns Hopkins University Press, 1993), as well as his earlier articles "The Decay of NASA's Technical Culture," *Space Policy* (November 1989): 301–310, and "Organizational Decline: NASA and the Life Cycle of Bureaus," *Public Administration Review* 51.4 (1991): 308–315.

54. For a reply to Downs, see Goodsell, *The Case for Bureaucracy*, chap. 6. For a reply to McCurdy, see Kay, *Can Democracies Fly in Space? The Challenge of Revitalizing the U.S. Space Program* (Westport, CT: Praeger, 1995), chap. 3.

55. Herbert A. Simon, *Administrative Behavior: A Study of Decision-Making Processes in Administrative Organization*, 3rd ed. (New York: Free Press, 1976), chap. 12.

56. Robert K. Merton, *Social Theory and Social Structure* (Glencoe, IL: Free Press, 1957).

57. Downs, *Inside Bureaucracy*. See also Merton, *Social Theory and Social Structure*; and Aaron Wildavsky, *Speaking Truth to Power: The Art and Craft of Policy Analysis* (Boston: Little, Brown, 1979), particularly pt. 1.

58. See note 37, as well as U.S. Congress, Office of Technology Assessment, *Starpower: The U.S. and the International Quest for Fusion Energy* (Washington: OTA, 1987).

59. W. D. Kay, "Policy-Making vs. Technology-Making: The Political Problems of Space R&D" (paper presented at the International Astronautical Federation Congress, Beijing, China, 1996), and "Does Hardware Vote? or Can Democracies Develop Expensive Large-Scale Technology?" (paper presented at the annual meeting of the American Political Science Association, Chicago, IL, 1995).

60. Eleanor Ehrnberg and Stefan Jaconsson, "Technological Discontinuities and Incumbents' Performance: An Analytical Framework," in *Systems of Innovation: Technologies, Institutions, and Organizations*, ed. Charles Edquist, 318–341 (London: Pinter, 1997); Bo Carlsson and Staffon Jacobsson, "Diversity

Creation and Technical Systems: A Technology Policy Perspective," in *Systems of Innovation: Technologies, Institutions, and Organizations*, ed. Charles Edquist, 266–294 (London: Pinter, 1997); W. Brian Arthur, "Competing Technologies, Increasing Returns, and Lock-In by Historical Events," *Economic Journal* 99.1 (1989): 116–131.

61. Goodsell, *The Case for Bureaucracy*.

62. Robert M. Wiseman and Philip Bromiley, "Toward a Model of Risk in Declining Organizations: An Empirical Examination of Risk, Performance, and Decline," *Organization Science* 7.5 (1996): 524–541.

63. See, for example, Clayton M. Christensen, *The Innovator's Dilemma: When New Technologies Cause Great Firms to Fail* (Boston: Harvard Business School Press, 1997); Michael Hammer and James Champy, *Reengineering the Corporation: A Manifesto for Business Revolution* (New York: Harper Business, 1993); David P. Angel, *Restructuring for Innovation: The Remaking of the U.S. Semiconductor Industry* (New York: Guilford, 1994).

64. Christensen, *The Innovator's Dilemma*; see also his "Exploring the Limits of the Technology S-Curve," *Productions and Operations Management* 1 (1992): 334–366; and Richard S. Rosenbloom and Clayton M. Christensen, "Technological Discontinuities, Organizational Capabilities, and Strategic Commitments," *Industrial and Corporate Change* 3 (1994): 655–685.

65. U.S. Congress, Office of Technology Assessment, *The Fusion Energy Program: The Role of TPX and Alternative Concepts*, background paper (Washington, DC: OTA, 1995).

66. See, for example, Deborah F. Crown and Joseph G. Rosse, "Yours, Mine, and Ours: Facilitating Group Productivity through the Integration of Individual and Group Goals," *Organizational Behavior and Human Decision Processes* 64.2 (1995): 138–150.

67. See Daniel J. Kevles, *The Physicists: The History of a Scientific Community in Modern America* (Cambridge, MA: Harvard University Press, 1995); and Adrienne Kolb and Lillian Hoddeson, "The Mirage of the 'World Accelerator for World Peace and the Origins of the SSC, 1953–1983," *HSPS: Historical Studies in the Physical and Biological Sciences* 24 (1993): 100–120.

68. Charles C. Alexander, *Holding the Line: The Eisenhower Era, 1952–1961* (Bloomington: Indiana University Press, 1975), 13. See also Richard Shenkman, *Presidential Ambition: Gaining Power at Any Cost* (New York: HarperCollins, 2000).

69. Some care must be taken here, since in many cases there may not be any direct information on the personal goals of individual decision makers. Sometimes, however, it is possible to make reasonable inferences on the basis of what is known about that person.

70. As will be seen (particularly in chapters 7 and 8), such political shifts can have an impact even when they do not mention space specifically.

71. Lester M. Salamon, "The Question of Goals," in *Federal Reorganization: What Have We Learned?* ed. Peter Szanton, 80 (Chatham, NJ: Chatham House, 1981).

72. Goodsell, *The Case for Bureaucracy*, chap. 7.

73. For a general discussion of reorganization, see Peter Szanton, ed., *Federal Reorganization: What Have We Learned?* (Chatham, NJ: Chatham House, 1981); B. Guy Peters, "Government Reorganization: A Theoretical Analysis," *International Political Science Review* 13.2 (1992): 199–217.

74. Peter Szanton, "So You Want to Reorganize the Government?" in *Federal Reorganization: What Have We Learned?* ed. Peter Szanton, 1–24 (Chatham, NJ: Chatham House, 1981). See also James Conant, "Reorganization and the Bottom Line," *Public Administration Review* 46.1 (1986): 48–56; George W. Downs and Patrick D. Larkey, *The Search for Government Efficiency: From Hubris to Helplessness* (New York: Random House, 1986), especially 184–190.

75. Andrew Lawler, "Agency Merger Plan Faces High Hurdles," *Science*, 31 March 1995, 1900; "Department of Science Still Evolving," *Science*, 14 April 1995, 191; Robert M. White, "Rekindling the Flame," *Technology Review* 98.4 (1995): 69; and David Hanson, "Congress Opens Science Department Hearings," *Chemical and Engineering News,* 10 July 1995, 22.

76. Goodsell, *The Case for Bureaucracy*, chap. 7.

77. Another variation of this phenomenon is a political leader attempting to place a particular individual in charge of a specific project. It has been suggested, for example, that President Johnson sought to transfer part of the Federal Aviation Administration's programs to the Department of Defense solely for the purpose of placing development of the Supersonic Transport under Robert McNamara's direction. See Szanton, "So You Want to Reorganize the Government?"

78. Radin and Hawley, *The Politics of Federal Reorganization*.

79. Szanton, "So You Want to Reorganize the Government?"

80. Radin and Hawley, *The Politics of Federal Reorganization*. See also Szanton, "So You Want to Reorganize the Government?"

81. See, for example, Charles Edquist and Björn Johnson, "Institutions and Organizations in Systems of Innovation," in *Systems of Innovation: Technologies, Institution, and Organizations*, ed. Charles Edquist, 41–63 (London: Pinter, 1997).

Chapter 3

1. For an account of early rocket research, see Roger Launius, "Prelude to the Space Age," in *Exploring the Unknown: Selected Documents in the History of the U.S. Civil Space Program*, vol. 1, *Organizing for Exploration*, ed. John M. Logsdon, 1–21 (Washington, DC: NASA, 1995); Howard E. McCurdy, *Space and the American Imagination* (Washington, DC: Smithsonian Institution Press, 1997), chap. 1; Eugene M. Emme, ed., *The History of Rocket Technology* (Detroit: Wayne State University Press, 1964); David Baker, *The Rocket: The History and Development of Rocket and Missile Technology* (New York: Crown Books, 1978); and Frank H. Winter, *Rockets into Space* (Cambridge, MA: Harvard University Press, 1990).

2. For more on the authors and the works themselves, see John Clute and Peter Nichols, *The Encyclopedia of Science Fiction* (New York: St. Martin's, 1995). The Hale story is reprinted in John M. Logsdon, ed., *Exploring the*

NOTES TO CHAPTER 3

Unknown: Selected Documents in the History of the U.S. Civil Service Program, vol. 1, *Organizing for Exploration* (Washington, DC: NASA, 1995), 23–55.

3. See (among many others) Richard Hutton, *The Cosmic Chase* (New York: Mentor, 1981), chap. 2; Tom Crouch, "'To Fly to the Moon': Cosmic Voyaging in Fact and Fiction from Lucian to Sputnik," in *Science Fiction and Space Futures, Past and Present*, ed. Eugene M. Emme, 7–26 (San Diego: Univelt, 1982); Launius, "Prelude to the Space Age," and *NASA: A History of the U.S. Civil Space Program* (Malabar, FL: Krieger, 1994), chap. 1; and McCurdy, *Space and the American Imagination*, chap. 1. Recently, science fiction publishers have sought to use this link to real-world spaceflight as a marketing technique. The cover of a reissued edition of Robert A. Heinlein's classic juvenile novel *Rocket Ship Galileo* (1947; New York: Ace Books), for example, proclaims that it was "the classic moon-flight novel that inspired modern astronautics."

4. Frederick I. Ordway III, "Evolution of Space Fiction in Film," in *History of Rocketry and Astronautics: AAS History Series*, vol. 11, ed. Roger D. Launius, 13–29 (San Diego: American Astronautical Society, 1994). It should be noted that some of the film characters of this period were in search of personal gain. The inventor of a spaceship in the Fritz Lang classic *Frau im Mond* (1929), for example, hoped to find gold on the moon. In addition, while the scientist who builds a spaceship to travel to Mars in *Just Imagine* wants only to travel where "mankind has never been," the pilot of the ship is primarily interested in becoming "distinguished" enough to marry his true love.

5. Quoted in A. A. Kosmodemjansky, "Konstantin Eduardovich Tsiolkovsky and the Present Times," in *History of Rocketry and Astronautics: AAS History Series*, vol. 11, ed. Roger D. Launius, see especially 173 (San Diego: American Astronautical Society, 1994). Goddard may well have shared these sentiments, but after the public ridicule that greeted his early work (including a scathing commentary in the *New York Times*), he would have never expressed them publicly.

6. Launius, "Prelude to the Space Age"; McCurdy, *Space and the American Imagination*, chap. 1. See also Frank H. Winter, *Prelude to the Space Age: The Rocket Societies, 1924–1940* (Washington, DC: Smithsonian Institution Press, 1983).

7. In fact, this feature of space travel stories continued almost into the 1950s. In Heinlein's *Rocket Ship Galileo*, a group of boys help an intrepid and resourceful inventor build a "cheap" single-stage spacecraft (out of a converted "mail rocket") that uses "super-heated zinc" (mercury would have been too expensive). There was, however, one exception to the nongovernment norm in science fiction. In 1946, a writer named Philip Klass (using the pen name William Tenn) published a story, "Alexander the Bait" in Astounding Science Fiction, that portrayed a "moon race" between space programs sponsored by governments (the Canadians win!). See William Tenn, "Alexander the Bait," *The Square Root of Man* (New York: Del Ray, 1968), 13–31.

8. This partnership would eventually lead to the establishment of the Jet Propulsion Laboratory in 1943. See Clayton R. Koppes, *JPL and the American Space Program: A History of the Jet Propulsion Laboratory* (New Haven, CT: Yale University Press, 1982).

9. An interesting comparison can be made in this regard between rocket research and other scientific fields. The physicist Ernest Lawrence (the inventor of the cyclotron), for example, was relentless in his pursuit of monetary support from any source. At one point, he was able to secure a grant from the National Institute of Health, claiming—correctly, as it turned out—that the radiation from his device could be used in treating some forms of cancer (he had determined this through experiments on his mother, which, fortunately had been overseen by his brother, a physician). See Herbert Childs, *An American Genius: The Life of Ernest Orlando Lawrence* (New York: Dutton, 1968); and Daniel J. Kevles, *The Physicists: The History of a Scientific Community in Modern America* (Cambridge, MA: Harvard University Press, 1995).

10. Louis Ferleger and William Lazonick, "The Managerial Revolution and the Developmental State: The Case of U.S. Agriculture," *Business and Economic History* 22.2 (1993): 67–98; W. D. Kay, Vicki L. Golich, and Thomas E. Pinelli, "U.S. Science and Technology Policy and the Production of Federally Funded Aeronautical Research and Technology," in *Knowledge Diffusion in the U.S. Aircraft Industry: Perspectives, Findings, and Improvement*, ed. Thomas E. Pinelli, 85–132 (Norwood, NJ: Ablex, 1997).

11. See Launius, "Prelude to the Space Age," 7, 10; and Koppes, *JPL and the American Space Program*.

12. The team had designated it the A-4. Its name was changed when Adolf Hitler decided to use it against London and other allied cities in retaliation (the V stood for vengeance) for bombing raids in Germany.

13. Indeed, the issue of the rocket team's (particularly von Braun's) complicity in war crimes—and the possibility that American military officials covered up the fact in order to get him into the United States—was never fully resolved during his lifetime. See Dennis Piszkiewicz, *Wernher von Braun: The Man who Sold the Moon* (Westport, CT: Praeger, 1998), particularly chaps. 1–3; and T. A. Heppenheimer, *Countdown: A History of Spaceflight* (New York: Wiley, 1997), 26–27; and William E. Burrows, *This New Ocean: The Story of the First Space Age* (New York: Modern Library, 1999), 102–106.

14. Quoted in Frederick I. Ordway and Mitchell R. Sharpe, *The Rocket Team* (New York: Crowell, 1979), 30.

15. Quoted in Piszkiewicz, *Wernher von Braun*, 27.

16. See the discussion in ibid., passim; Wayne Biddle, "Science, Morality, and the V-2," *New York Times* (October 2, 1992), A31; and Launius, "Prelude to the Space Age," 12.

17. Quoted in Piszkiewicz, *Wernher von Braun*, 33.

18. Ibid.

19. Burrows, *This New Ocean*, 105–106.

20. See Walter Dornberger, *V-2: The Nazi Rocket Weapon* (New York: Viking, 1954); and Michael Neufeld, "Hitler, the V-2, and the Battle for Priority," *Journal of Military History* (July 1993): 511–538.

21. Michael Burleigh, *The Third Reich: A New History* (New York: Hill and Wang, 2000), 747.

NOTES TO CHAPTER 3

22. Quoted in Burrows, *This New Ocean*, 106.

23. Actually, it is not altogether clear exactly what caused Hitler to change his mind about the program, von Braun's enthusiasm or his own excitement after seeing the film. Speer's account of the meeting (cited in ibid.), for example, does not mention von Braun at all. Still, even if Hitler did essentially talk himself into supporting the V-2, there is no evidence that von Braun sought to dissuade him.

24. Quoted in William Sims Bainbridge, *The Spaceflight Revolution: A Sociological Study* (New York: Wiley, 1976), 67.

25. Described in G. Pascal Zachary, *Endless Frontier: Vannevar Bush, Engineer of the American Century* (New York: Free Press, 1997), 318.

26. See Heppenheimer, *Countdown*, 47.

27. The quote is from Zachary, *Endless Frontier* (316), which also suggests (318) that Bush's objections may have come as much from personal prejudice as from technical reasoning.

28. Not to mention smaller and lighter nuclear weapons.

29. Arthur C. Clarke, "Extra-Terrestrial Relays: Can Rocket Stations Give World-Wide Radio Coverage?" *Wireless World* (October 1945): 305–308.

30. American Rocket Society Space Flight Committee, "On the Utility of an Artificial Unmanned Earth Satellite: A Proposal to the National Science Foundation." *Jet Propulsion* 25 (February 1955): 71–78; reprinted in Logsdon, ed., *Exploring the Unknown*, 281–294.

31. For a discussion of how Pearl Harbor affected the U.S. intelligence community, see Dwayne A. Day, John M. Logsdon, and Brian Lattell, Introduction to *Eye in the Sky: The Story of the CORONA Spy Satellites*, ed. Dwayne A. Day, John M. Logsdon, and Brian Latell,see especially 2–3 (Washington, DC: Smithsonian Institution Press, 1998).

32. Who would later serve as Eisenhower's—and the nation's—first presidential science advisor.

33. The Killian Report, "Meeting the Threat of a Surprise Attack," is discussed in James R. Killian Jr., *Sputniks, Scientists, and Eisenhower: A Memoir of the First Special Assistant to the President for Science and Technology* (Cambridge, MA: MIT Press, 1997), 72–79. See also Dwayne A. Day, "A Strategy For Space: Donald Quarles, the CIA, and the US Scientific Satellite Program," *Spaceflight* 38 (1996): 308–312, and "A Strategy for Reconnaissance: Dwight D. Eisenhower and Freedom of Space," in *Eye in the Sky: The Story of the CORONA Spy Satellites*, ed. Dwayne A. Day, John M. Logsdon, and Brian Laqttell, 119-142 (Washington, DC: Smithsonian Institution Press, 1998).

34. See R. Cargill Hall, "Postwar Strategic Reconnaissance and the Genesis of CORONA," in *Eye in the Sky: The Story of the CORONA Spy Satellites*, ed. Dwayne A. Day, John M. Logsden, and Brian Laqttell, 86–118, (Washington, DC: Smithsonian Institution Press, 1998) and "Origins of U.S. Space Policy: Eisenhower, Open Skies, and Freedom of Space," in *Exploring the Unknown: Selected Documents in the History of the U.S. Civil Service Program*, vol. 1, *Organizing for Exploration*, ed. John M. Logsdon 213–229 (Washington, DC: NASA, 1995); and Charles C. Alexander, *Holding the Line: The Eisenhower Era, 1952–1961* (Bloomington: Indiana University Press, 1975), 96–97.

35. Hall, "Postwar Strategic Reconnaissance." See also Tom D. Crouch, *The Eagle Aloft: Two Centuries of the Balloon in America* (Washington, DC: Smithsonian Institution Press, 1983), 644–649.

36. Douglas Aircraft Company, "Preliminary Designs of an Experimental World-Circling Spaceship," Report No. SM-11827, 2 May 1946, in, *Exploring the Unknown: Selected Documents in the History of the U.S. Civil Service Program*, vol. 1, *Organizing for Exploration*, ed. John M. Logsdon 236–244 (Washington, DC: NASA, 1995). See also J. E. Ripp, R. M. Salter, and W. S. Wernher et. al., "The Utility for a Satellite Vehicle for Reconnaissance" RAND Report R-217, April 1951, in *Exploring the Unknown: Selected Documents in the History of the U.S. Civil Service Program*, vol. 1, *Organizing for Exploration*, ed. John M. Logsdon 245–261 (Washington, DC: NASA, 1995).

37. Quoted in McCurdy, *Space and the American Imagination*, 67.

38. Naturally, both the United States and the USSR began during the 1960s to develop antisatellite weapons; see the discussion in chap. 7.

39. One must be careful, however, not to take such a claim too far. A staunch fiscal conservative, President Eisenhower was highly suspicious of what he regarded as large, costly government programs. Thus, although the satellite program was accorded a far higher priority than it had been before—and higher than if it had been regarded primarily as simply a scientific program—it was still considered less important than other high-tech military programs (such as the ICBM), and had to operate under strict spending limits. See Logsdon, "Origins of U.S. Space Policy," 224.

40. In fact, the first mission of the first successful spy satellite (part of the CORONA program), returned more data than all the U-2 flights combined. See Albert D. Wheelon, "CORONA: A Triumph of American Technology" (29–47) and Dwayne A. Day, "The Development and Improvement of the CORONA Satellite," in *Eye in the Sky: The Story of the CORONA Spy Satellites*, ed. Dwayne A. Day, John M. Logsdon, and Brian Lattell, 48–85 (Washington, DC: Smithsonian Institution Press, 1998).

41. It would later be renamed Project Corona.

42. See William E. Burrows, *Deep Black: Space Espionage and National Security* (New York: Random House, 1987), vii. See also the discussion in Day, Logsdon, and Lattell, "Introduction," n. 1.

43. In addition, the documents related to the program were not declassified until the 1990s. See Day, Logsdon, and Lattell, "Introduction."

44. McCurdy, *Space and the American Imagination*, chap. 2.

45. For more on IGY, see R. G. Fleagle, "From the International Geophysical Year to Global Change," *Reviews of Geophysics* 30 (1992): 305–313; and Heppenheimer, *Countdown*, 92–93.

46. Wernher von Braun, "A Minimum Satellite Vehicle: Based on Components Available from Missile Developments of the Army Ordinance Corps," 15 September 1954, in *Exploring the Unknown: Selected Documents in the History of the U.S. Civil Service*, vol. 1, *Organizing for Exploration*, ed. John M. Logsdon, 274–281 (Washington, DC: NASA, 1995). American Rocket Society Space Flight Committee, "On the Utility of an Artificial Unmanned Earth Satellite"; and S. F.

Singer, "Studies of a Minimum Orbital Unmanned Satellite of the Earth (MOUSE)," *Astronautica Acta* 1 (1955): 171–184, reprinted in Logsdon, ed., *Exploring the Unknown*, 1:314–324.

47. In fact, this would not be the first time that official silence would appear conspicuous. During World War II, the sudden disappearance of any references to nuclear fission in popular books and magazines led some (although apparently not anyone in Germany or Japan) to suspect that the United States must have been working on an atomic bomb. See Herbert N. Forestel, *Secret Science: Federal Control of Science and Technology in America* (Westport, CT: Praeger, 1993).

48. See, for example, Hall, "Origins of U.S. Space Policy"; Walter A. McDougall, . . . *the Heavens and the Earth: A Political History of the Space Age* (New York: Basic Books, 1985); and Dwayne A. Day, "A Strategy for Space," 308–312, and "A Strategy for Reconnaissance," 119–142.

49. Day, "A Strategy for Reconnaissance," 119–120.

50. White House press release, July 29, 1955, NASA History Office (hereafter NHO), also in Logsdon, ed., *Exploring the Unknown*, 1:200–201.

51. U.S. National Security Council, *Draft Statement of Policy on U.S. Scientific Satellite Program: General Considerations*, NSC 5520 (20 May 1955), in NHO, also in Logsdon, ed., *Exploring the Unknown*, 1:308–313.

52. U.S. National Security Council, *Draft Statement*, annex A, "Technical Annex." This section of the report also notes, however, that these scientific observations could also have great military value. Additional knowledge about movement through the upper atmosphere, for example, would help contribute to the ICBM program, as well as to the development of antimissile countermeasures.

53. The document also contains numerous warnings that the IGY satellite not interfering with "other programs" (unnamed, but presumably referring to efforts like the WS-117L).

54. Memorandum from Percival Brundage, director, Bureau of the Budget, to the president, "Project Vanguard," 30 April 1957, in *Exploring the Unknown: Selected Documents in the History of the U.S. Civil Space Program*, vol. 2, *External Relationships*, ed. John M. Logsdon, 277–280 (Washington, DC: NASA, 1996). Significantly, this memo also indicates that a significant amount of money (over $2 million) was contributed to this "scientific" program by the Central Intelligence Agency.

55. Which, as noted earlier, has only been fully understood within the past decade or so.

56. See, for example, Charles Lindbergh's foreword to *Vanguard: A History*, by Constance McLaughlin Green and Milton Lomask (Washington, DC: NASA, 1970), x–xi; Richard Hutton, *The Cosmic Chase* (New York: Mentor, 1981), chap. 3; the "disaster" characterization comes from McDougall, . . . *the Heavens and the Earth*, 122.

57. David Halberstam, *The Fifties* (New York: Villard, 1993), 623.

58. For more details about the selection process, see Constance McLaughlin Green and Milton Lomask, *Vanguard: A History* (Washington, DC:

NASA, 1970), especially 41–55; and McDougall, . . . *the Heavens and the Earth*, chap. 5.

59. Which would eventually become the Goddard Space Center.

60. The air force did submit what has been characterized as a "half-hearted proposal," which does not seem to have been a serious contender. See Dwayne A. Day, "Invitation to Struggle: The History of Civilian-Military Relations in Space" in *Exploring the Unknown: Selected Documents in the History of the U.S. Civil Space Program*, vol. 2, *External Relationships*, ed. John M. Logsdon, especially 342 (Washington, DC: NASA, 1996).

61. This claim was almost certainly correct: the von Braun team successfully launched the first U.S. satellite, Explorer I, just 90 days after receiving approval.

62. A more complete discussion can be found in Harvey M. Sapolsky, *Science and the Navy: The History of the Office of Naval Research* (Princeton, NJ: Princeton University Press, 1990).

63. Halberstam, *The Fifties*, 623.

64. A.V. Grosse to Donald A. Quarles, "A Report on the Present Status of the Satellite Problem," 25 August 1953, in Logsdon, ed., *Exploring the Unknown*, 1:268.

65. Von Braun, "A Minimum Satellite Vehicle."

66. U.S. National Security Council, *Draft Statement*.

67. Memorandum from Nelson A. Rockefeller to James S. Lay Jr., executive secretary, National Security Council, 17 May 1955, NSC 5520, annex B (see Logsdon, ed., *Exploring the Unknown*, 1:312–313).

68. See, for example, the letter from Allen W. Dulles, director of Central Intelligence, to Donald Quarles, deputy secretary of Defense, 5 July 1957, in Logsdon, ed., *Exploring the Unknown*, 1:329. Evidently, Alexander Nesmsyanov, president of the Soviet Academy, was not at all reticent about commenting on his country's satellite program. *Science* magazine had reported one month earlier that he had been quoted in Pravda as stating that the USSR had essentially solved all of the technical problems related to an artificial earth satellite. See James R. Killian Jr., *Sputniks, Scientists, and Eisenhower: A Memoir of the First Special Assistant to the President for Science and Technology* (Cambridge, MA: MIT Press, 1997), 2.

69. William E. Burrows, *Deep Black: Space Espionage and National Security* (New York: Random House, 1987), 90–92. Apparently, the one aspect of the Soviet program that U.S. intelligence got wrong was the power of its rockets. Eisenhower later admitted to being genuinely surprised by the weight of *Sputnik I* (184 pounds). See Dwight D. Eisenhower, *The White House Years: Waging Peace*. (Garden City, NY: Doubleday, 1965), 205.

70. Burrows, *Deep Black*; Hall, "Origins of U.S. Space Policy"; David Callahan and Fred I. Greenstein, "The Reluctant Racer: Eisenhower and U.S. Space Policy," in *Spaceflight and the Myth of Presidential Leadership*, ed. Roger D. Launius and Howard E. McCurdy, 15–50 (Urbana: University of Illinois, 1997).

71. Killian, *Sputniks, Scientists, and Eisenhower*, 10; Richard Nixon, *RN: The Memoirs of Richard Nixon* (New York: Simon and Schuster, 1980), 428–429.

72. Halberstam, *The Fifties*, 624.

73. For a review of this literature, see Steven Rabe, "Eisenhower Revisionism: A Decade of Scholarship," *Diplomatic History* 17 (1993): 97–115; and Stephen E. Ambrose, *Eisenhower: The President* (New York: Simon and Schuster, 1984), chap. 27.

74. For the most recent statement of this charge, see Edwin Diamond, "Sputnik," *American Heritage* 48.6 (1997): 84–93.

75. As he did, for example, in "The President's News Conference of January 26, 1960," *Public Papers of the President: Dwight David Eisenhower, 1960* (Washington, DC: Government Printing Office, 1960), 127.

Chapter 4

1. *Sputnik III*, launched the following May, weighed one and a half tons.

2. It is now known (although it was also suspected at the time) that in reality the Russian space program was not as advanced as it appeared. Beset by far more technical difficulties and breakdowns than they revealed to the outside world, the Soviets more than once placed their personnel in great danger. In 1960, for example, dozens of scientists, technicians, and soldiers were killed in a launch pad explosion, reportedly the result of political pressure to launch a probe so that Premier Nikita Khrushchev could announce it during a trip to New York. Similar problems and near catastrophes would occur later during the USSR's manned flights (see James Oberg, *Red Star in Orbit: The Inside Story of Soviet Failures and Triumphs in Space* [New York: Random House, 1981], especially chap. 3). Nevertheless, as far as many American officials and the general public were able to tell, the Soviet "lead" in space seemed very great indeed.

3. For a review of early Soviet space achievements, see Oberg, *Red Star in Orbit*; John Noble Wilford, "First into Space, then the Race," in *Anatomy of the Soviet Union*, ed. Harrison E. Salisbury, 342–357 (London: Nelson, 1967); and Ronald D. Humble, *The Soviet Space Program* (London: Routledge, 1988).

4. See Constance McLaughlin Green and Milton Lomask, *Vanguard: A History* (Washington, DC: NASA, 1970).

5. Since, in the administration's view, *Sputnik* had established the "freedom of space" so important to the president, there was no longer any need for the U.S. satellite program to stay "purely" scientific. Von Braun was given approval to begin work on *Explorer I* in late October 1957.

6. "The President's News Conference of October 9, 1957," *Public Papers of the Presidents: Dwight D. Eisenhower, 1957* (Washington, DC: GPO, 1958), 719–732.

7. "Radio and Television Address to the American People on Science and National Security, November 7, 1957" (789–799), and "Radio and Television Address to the American People on 'Our Future Security,' November 13, 1957" (809–816), in *Public Papers of the Presidents: Dwight D. Eisenhower, 1957* (Washington, DC: GPO, 1958).

8. Personal letter from Dwight David Eisenhower to Lloyd S. Swenson, August 5, 1965, in NASA Historical Reference Collection (hereafter NHRC).

9. Ibid.

10. A point made quite persuasively in Rip Bulkeley, *The Sputniks Crisis and Early United States Space Policy* (Bloomington: Indiana University Press, 1990).

11. Stephen E. Ambrose, *Eisenhower: The President* (New York: Simon and Schuster, 1984), 425.

12. Memorandum of Conference with the President, 8 October, 1957. Copies in NHRC and Dwight D. Eisenhower Presidential Library (hereafter DEL).

13. "Memorandum: Discussion at the 339th Meeting of the National Security Council, October 10, 1957," in NHRC and DEL.

14. Quoted in Donald W. Cox, *The Space Race* (Philadelphia: Chilton Books, 1962), 137.

15. Reproduced in Eugene M. Emme, "Presidents and Space," in *Between Sputnik and the Shuttle: New Perspectives on American Astronautics*, ed. Frederick C. Curant III, 5–138, 21 (San Diego: Univelt, 1981).

16. George E. Price, "Arguing the Case for Being Panicky," *Life*, 18 November 1957, 125.

17. Quoted in R. Cargill Hall, "Origins of U.S. Space Policy: Eisenhower, Open Skies, and Freedom of Space," in *Exploring the Unknown: Selected Documents in the History of the U.S. Civil Space Program*, vol. 1, *Organizing for Exploration*, ed. John M. Logsdon, 213–225 (Washington, DC: NASA, 1995).

18. Quoted in Lyndon Baines Johnson, *The Vantage Point: Perspectives on the Presidency, 1963–1969* (New York: Holt, Rinehart, and Winston, 1971), 273. See also Richard Hutton, *The Cosmic Chase* (New York: Mentor, 1981), 37.

19. These charges are discussed in Walter A. McDougall, . . . *the Heavens and the Earth: A Political History of the Space Age* (New York: Basic Books), 1985, chap. 6.

20. Richard Nixon, *RN: The Memoirs of Richard Nixon* (New York: Simon and Schuster, 1980), 428–429. There is some evidence, however, that this statement represents a bit of revisionism on Nixon's part. The minutes of the October 10, 1957, NSC meeting (see note 13) state that the vice president expressed "full support" for Eisenhower's position on *Sputnik* and the U.S. rocket program. Of course, vice presidents are expected to support the administration of which they are a part; it is always possible that Nixon expressed his "frustrations" with the president in private.

21. See, for example, David Halberstam, *The Fifties* (New York: Villard, 1993), 625; McDougall, . . . *the Heavens and the Earth*, 146–148.

22. U.S. Congress, House of Representatives, *Toward the Endless Frontier: History of the Committee on Science and Technology, 1959–1979* (Washington, DC: GPO, 1990), 2–4.

23. See also interview with Gerald W. Siegel, staff worker for (then) Senator Lyndon B. Johnson, by Jay Holmes of the NASA Historical Staff, 25 June 1968, 2, in NHRC.

24. See, for example, Roger D. Launius, "Eisenhower, Sputnik, and the Creation of NASA: Technological Elites and the Public Policy Agenda," *Prologue* 28.2 (1996): 127–143.

25. Charles C. Alexander, *Holding the Line: The Eisenhower Era, 1952–1961* (Bloomington: Indiana University Press, 1975), 101. See also John Kenneth Galbraith, *The Affluent Society* (Boston: Houghton Mifflin, 1958).

26. Ambrose, *Eisenhower*, 424.

27. For a useful summary of these feelings, see Roger D. Launius, "Prelude to the Space Age," in *Exploring the Unknown: Selected Documents in the History of the U.S. Civil Space Program*, vol. 1, *Organizing for Exploration*, ed. John M. Logsdon, 1–22 (Washington, DC: NASA, 1995).

28. James R. Killian Jr., *Sputniks, Scientists, and Eisenhower: A Memoir of the First Special Assistant to the President for Science and Technology* (Cambridge, MA: MIT Press, 1997), 2.

29. Tom Wolfe, *The Right Stuff* (New York: Bantam Books, 1979).

30. "Announcement of the First Satellite," *Pravda*, October 5, 1957, quoted in John M. Logsdon, ed., *Exploring the Unknown: Selected Documents in the History of the U.S. Civil Space Program*, vol. 1, *Organizing for Exploration* (Washington, DC: NASA, 1995), 329–330.

31. *Outline of the History of the USSR*, trans. George H. Hanna (Moscow: Foreign Languages Publishing House, 1960), 361.

32. Quoted in McDougall, . . . *the Heavens and the Earth*, 328.

33. Ibid., 8.

34. "Annual Message to the Congress on the State of the Union," *Public Papers of the Presidents: Dwight D. Eisenhower, 1958* (Washington, DC: GPO, 1959), 2–3.

35. *Congressional Record*, House, 2 June 1958, 9919–9920.

36. Bernard Holland, "Van Cliburn: Man behind the Contest," *New York Times*, 27 March 1989, C13.

37. See, for example, Max Frankel, "U.S. Pianist, 23, Wins Soviet Contest," *New York Times*, 14 April 1958, A1.

38. See, for example, Edward A. Purcell Jr., *The Crisis of Democratic Theory: Scientific Naturalism and the Problem of Value* (Lexington: University Press of Kentucky, 1973).

39. "The President's News Conference of October 9, 1957," *Public Papers of the Presidents: Dwight D. Eisenhower, 1957* (Washington, DC: GPO, 1958), 722.

40. "The President's News Conference at Augusta, Georgia, October 22, 1959," *Public Papers of the President: Dwight D. Eisenhower, 1959* (Washington, DC: GPO, 1960), 734.

41. "The President's News Conference of July 29, 1959," *Public Papers of the President: Dwight D. Eisenhower, 1959* (Washington, DC: GPO, 1960), 555.

42. Roger D. Launius, *NASA: A History of the U.S. Civil Space Program* (Malabar, FL: Krieger, 1994), 24–25; and Edwin Diamond, "Sputnik," *American Heritage* 48.6 (1997): 84–93.

43. "Memorandum: Discussion at the 339th Meeting of the National Security Council, October 10, 1957," in NHRC and DEL.

44. "Memorandum of Conference with the President, 8 October, 1957," in NHRC and DEL.

45. See, for example, Launius, "Eisenhower, Sputnik, and the Creation of NASA," as well as the discussion in the preceding chapter.

46. W. D. Kay, interview with George Brown, 10 July 1997. See also Daniel J. Kevles, *The Physicists: The History of a Scientific Community in Modern America* (Cambridge, MA: Harvard University Press, 1995), chap. 23.

47. U.S. Congress, House of Representatives, *Independent Offices: Hearings before a Subcommittee of the Committee on Appropriations*, 86th Cong., 1st sess., 1959, 322.

48. W. D. Kay, "Congressional Decision-Making and Long-Term Technological Development: The Case of Nuclear Fusion," in *Science, Technology, and Politics: Policy Analysis in Congress*, ed. Gary Bryner, 87–106 (Boulder, CO: Westview, 1992).

49. With some important exceptions (see chapter 6).

50. Paul Forman, "Behind Quantum Electronics: National Security as Basis for Physical Research in the United States, 1940–1960 (Part I)," *HSPS: Historical Studies in Physical and Biological Studies* 15.1 (1985): 149–229; Kevles, *The Physicists*; and Stuart M. Feffer, "Atoms, Cancer, and Politics: Supporting Atomic Science at the University of Chicago, 1944–1950," *HSPS: Historical Studies in the Physical and Biological Sciences* 22.2 (1982): 233–261.

51. Daniel S. Greenberg, *The Politics of Pure Science* (New York: New American Library, 1967), 215 fn.

52. See, for example, Halberstam, *The Fifties*, especially 9–40.

53. Alexander, *Holding the Line*, 242.

54. Eileen Galloway, "U.S. Congress and Outer Space," in *Between Sputnik and the Shuttle: New Perspectives on American Astronautics*, ed. Frederick C. Durant III, 139–160 (San Diego: Univelt, 1981); and U.S. Congress, House of Representatives, *Toward the Endless Frontier*, 1.

55. Johnson, *The Vantage Point*, 272–273.

56. Siegel interview.

57. Memo to Lyndon Baines Johnson from George E. Reedy, 17 October 1957, LBJ Presidential Library (hereafter LBJL).

58. McDougall, . . . *the Heavens and the Earth*, 148–149; and Bulkeley, *The Sputniks Crisis*, 186–187.

59. Memo to Johnson from Reedy, 17 October 1957, in LBJL.

60. Reported in Ambrose, *Eisenhower*, 430. See also the discussion in the last chapter.

61. U.S. Congress, Senate, *Inquiry into Satellite and Missile Programs: Hearings before the Preparedness Investigating Subcommittee of the Committee on Armed Services*, 85th Cong., 1st and 2nd sess. (25–27 November and 13–17 December 1957; 10–23 January 1958), 1958, 2–3.

62. Memo to Johnson from Reedy, 17 October 1957 (emphasis added), in LBJL.

63. Memo to Lyndon Baines Johnson from James Rowe, 15 November 1957, in LBJL.

64. Memo to Walter Jenkins from Glen Wilson, 4 April 1958, in LBJL.

65. "Statement of Democratic Leader Lyndon B. Johnson to the Meeting of Democratic Conference on January 7, 1958," in LBJL and NHRC.

66. Ibid.

67. Detailed accounts of the birth of NASA can be found in McDougall, . . . *the Heavens and the Earth*, chap. 7; Launius, *NASA*, chap. 3, and "Eisenhower, Sputnik, and the Creation of NASA"; and Killian, *Sputniks, Scientists, and Eisenhower*, 122–138. See also the documents reproduced in Logsdon, ed., *Exploring the Unknown*, 1:628–643.

68. Peter B. Dow, *Schoolhouse Politics: Lessons from the Sputnik Era* (Cambridge, MA: Harvard University Press, 1991).

69. Killian, *Sputniks, Scientists, and Eisenhower*. See also Bruce L. R. Smith, *The Advisors: Scientists and the Policy Process* (Washington, DC: Brookings Institution, 1992), chap. 8.

70. James M. Grimwood, *Project Mercury: A Chronology* (Washington, DC: NASA SP4001, 1963), 12. See also Bruce L. R. Smith, *American Science Policy since World War II* (Washington, DC: Brookings Institution, 1990), 55.

71. U.S. Congress, Office of Technology Assessment, *Federally Funded Research: Decision for a Decade* (Washington, DC: OTA, 1991), 112. Figures are in constant 1982 dollars.

72. Ibid., 114.

73. Galloway, "U.S. Congress and Outer Space"; and U.S. Congress, House of Representatives, *Toward the Endless Frontier*, 1.

74. See U.S. Congress, House of Representatives, *Toward the Endless Frontier*, 15–17.

75. Senate Committee on Government Operations, Staff Memorandum No. 85-1-73, 8 November 1957, copies in DEL and NHRC; and Launius, "Eisenhower, Sputnik, and the Creation of NASA."

76. U.S. Congress, House of Representatives, "The Problems of Congress in Formulating Outer Space Legislation,"*Astronautics and Space Exploration: Hearings before the Select Committee on Astronautics and Space Exploration*, 85th Cong., 2nd sess., 15–30 April and 1–12 and May 1958, 5–10; and Killian, *Sputniks, Scientists, and Eisenhower*, especially 30–32.

77. "Message from the President of the United States Relative to Space Science and Exploration," House Document No. 365, 85th Cong., 2nd sess., 2 April 1958.

78. U.S. Congress, House of Representatives, *Astronautics and Space Exploration*, 15–30 April and 1–12 May, 1958, 1–2.

79. U.S. Congress, Senate, *National Aeronautics and Space Act: Hearings before the Select Committee on Space and Astronautics*, 85th Cong., 2nd sess., 6–8 May 1958, 6.

80. *Congressional Record*, House, 2 June 1958, 9917.
81. Ibid., 9917–9918.
82. Ibid., 9918.
83. Ibid., 9921.
84. Ibid., 9926.
85. *Congressional Record*, Senate, 16 June 1958, 11293. See also Launius, *NASA*, 31.
86. Which, in fact, may well have been Johnson's intention. See Skolnikoff interview with Killian (1968?), 11, in NHRC.
87. See Killian, *Sputniks, Scientists, and Eisenhower*, 137.
88. "Statement by the President upon Signing the National Aeronautics and Space Act of 1958, July 29, 1958," *Public Papers of the President: Dwight David Eisenhower, 1958* (Washington, DC: GPO, 1959), 573.
89. J. D. Hunley., ed., *The Birth of NASA: The Diary of T. Keith Glennan* (Washington, DC: NASA SP-4105, 1993), 2.
90. See discussion in the next chapter.
91. Hunley, *The Birth of NASA*, 31. He does, however, acknowledge that there was a competition between the two countries.
92. U.S. Congress, House of Representatives, *1960 NASA Authorization: Hearings before the Committee on Science and Astronautics and Subcommittees 1, 2, 3, 4*, 86th Cong., 1st sess., 20–29 April and 4 May, 1959, 3–4; emphasis added.
93. U.S. President's Science Advisory Committee, *Introduction to Outer Space: An Explanatory Statement* (Washington, DC: GPO, 1958, 1–2.
94. National Security Council, *Preliminary U.S. Policy on Outer Space*, NSC 5814/1, 18 August 1958, in NHRC.
95. Ibid.
96. National Security Council, *Preliminary U.S. Policy on Outer Space*, NSC 5814, 20 June 1958, in Logsdon, ed., *Exploring the Unknown*, 1:353.
97. The practice here will be to adopt whatever phrasing was used at the time being examined. Thus, instead of the phrase *human spaceflight* currently employed by NASA, this section uses the term *manned spaceflight*, as will subsequent sections.
98. U.S. Congress, Senate, *Project Mercury: Man-in-Space Program of the National Aeronautics and Space Administration: Report of the Committee on Aeronautical and Space Sciences*, 86th Cong., 1st sess., 1 December 1959, 1.
99. Lloyd S. Swenson Jr., James M. Grimwood, and Charles C. Alexander, *This New Ocean: A History of Project Mercury*. NASA History Series (Washington, DC: NASA, 1966), 91; and Grimwood, *Project Mercury*, 13.
100. Hunley, *The Birth of NASA*, 13. See also U.S. Congress, House of Representatives, *United States Civilian Space Programs 1958–1978: Report Prepared for the Subcommittee on Space Science and Applications, Committee on Science and Technology*, 97th Cong., 1st sess., January 1981, 347. Responsibility for the man-in-space program was transferred to NASA by President Eisenhower (acting on advice from Killian) in August 1958 (see Swenson, Grimwood, and Alexander, *This New Ocean*,

101–102). On November 26, it was officially named Project Mercury (Grimwood, *Project Mercury*, 32).

101. Grimwood, *Project Mercury*, 31.
102. U.S. Congress, Senate, *Project Mercury*, 2.
103. President's Science Advisory Committee, "Report of the Ad-Hoc Committee on Man-in-Space," 16 December 1960, in Launius, *NASA*, 167; emphasis added.
104. National Aeronautics and Space Council, "U.S. Policy on Outer Space," 26 January 1960, in Logsdon, ed., *Exploring the Unknown*, 1:365.
105. On Eisenhower and Mercury, see Ambrose, *Eisenhower*, 591–592.
106. Swenson, Grimwood, and Alexander, *This New Ocean*, 306.
107. See, for example, "Difficult Decisions," *Science*, 1 November 1960, 3434; and "Too Much Speed in Space?" *Wall Street Journal*, 19 December 1960, 5. For a general discussion, see William David Compton, *Where No Man Has Gone Before: A History of Apollo Lunar Exploration Missions* (Washington, DC: NASA SP-4214, 1989), 16.
108. Swenson, Grimwood, and Alexander, *This New Ocean*, 304–305.
109. Launius, *NASA*, 40. For an especially entertaining look at this period, see Wolfe, *The Right Stuff*.
110. U.S. National Aeronautics and Space Administration, *Aeronautics and Space Report of the President: Fiscal Year 1995 Activities* (Washington, DC: NASA, 1996), app. A-1. It was widely believed that the USSR had had a number of failures as well, but simply did not acknowledge them.
111. See, for example, "Too Much Speed in Space?"; and William R. Corliss, *Scientific Satellites* (Washington, DC: NASA SP-133, 1967), chap. 2.
112. Oberg, *Red Star in Orbit*.
113. Swenson, Grimwood, and Alexander, *This New Ocean*, 307.
114. The major exception here, of course, is the so-called (and, as it later turned out, fictitious) "missile gap" with the USSR that the Eisenhower administration had allegedly permitted to "open up." This, however, was not necessarily associated in the public mind with space exploration.
115. Quoted in Courtney Sheldon, "The Space Challenge Issue," *Christian Science Monitor*, 10 August 1960, 7.
116. Democratic Advisory Council, *Position Paper on Space Research* (Washington, DC: Democratic Advisory Council, September 7, 1960), 3. Copies in NHRC and John F. Kennedy Presidential Library (hereafter JFKL).
117. Quoted in Jack Doherty, "Space in the 1960 Campaign," 21 July 1963, in NHRC.
118. "Inaugural Address of President John F. Kennedy," 20 January 1961, *Public Papers of the President: John F. Kennedy, 1961* (Washington, DC: GPO, 1962), 2.
119. "Annual Message to the Congress," 30 January 1961, *Public Papers of the President: John F. Kennedy, 1961* (Washington, DC: GPO, 1962), 26–27.
120. "Draft Paper on US–USSR Space Cooperation," 4 April 1961, in *Exploring the Unknown: Selected Documents in the History of the U.S. Civil Space*

Program, vol. 2, *External Relationships*, ed. John M. Logsdon, 143 (Washington, DC: NASA, 1997).

121. Launius, *NASA*, 57–58; and McDougall, . . . *the Heavens and the Earth*, chap. 17. For a more in-depth discussion, see W. D. Kay, "John F. Kennedy and the Two Faces of the U.S. Space Program, 1961–1963," *Presidential Studies Quarterly* 28.3 (1998): 573–586.

122. John F. Kennedy, "If the Soviets Control Space, They Can Control Earth," *Missiles and Rockets*, 10 October 1960, 12–13.

123. See Linda T. Krug, *Presidential Perspectives on Space Exploration: Guiding Metaphors from Eisenhower to Bush* (New York: Praeger, 1991), chap. 2.

124. Hugh Sidey, *John F. Kennedy: President* (New York: Atheneum, 1964), 118.

125. A practice he continued throughout the Johnson presidency. See chapters 5 and 6.

126. "Administrator's Presentation to the President," 21 March 1961, in NHRC. See also Webb's account of the meeting in W. Henry Lambright, *Powering Apollo: James E. Webb of NASA*, Johns Hopkins New Series in NASA History (Baltimore: Johns Hopkins University Press, 1995), 92.

127. In December 1960, Kennedy had suggested to Johnson that he take over as chair of the Space Council. Congress approved an amendment to the National Aeronautics and Space Act on April 20, 1961, by which time the issue had already been settled (see next chapter). See Johnson, *The Vantage Point*, 278.

128. At least not at that time. Today, there are a number of coordinating councils, covering a wide variety of policy areas, within the executive office of the president.

129. *Congressional Record*, House, 2 June 1958, 9927.

130. It is worth pointing out that even this "cut" budget was still just below $1 billion.

131. W. D. Kay, *Can Democracies Fly in Space? The Challenge of Revitalizing the U.S. Space Program* (Westport, CT: Praeger, 1995), 48–51.

132. See, for example, Hutton, *The Cosmic Chase*, 50–51.

133. Ambrose, *Eisenhower*, 433.

Chapter 5

1. Robert T. Nakamura and Frank Smallwood, *The Politics of Policy Implementation* (New York: St. Martin's, 1980).

2. For a discussion of the War on Poverty and its programs, see Michael L. Gillette, *Launching the War on Poverty: An Oral History* (London: Prentice-Hall, 1996); and Edward Ziegler and Jeannette Valentine, eds., *Project Head Start: A Legacy of the War on Poverty* (New York: Free Press, 1979). On Operation Desert Storm, see Bob Woodward, *The Commanders* (New York: Simon and Schuster, 1991); James K. Matthews, *So Many, So Much, So Far, So Fast: United States Transportation Command and Its Deployment for Operation Desert Shield/Desert Storm* (Washington, DC: Joint History Office, Office of the Chairman of the Joint Chiefs of Staff and the Research Center of U.S. Transportation Command, 1996).

3. Questions about how Gagarin had landed were to prove somewhat embarrassing for the Soviets when they claimed world records for the flight, since the rules of the Federation Aeronautique Internationale (the organization that certifies such records) require that a pilot take off and land inside the same vehicle. See James Oberg, *Red Star in Orbit: The Inside Story of Soviet Failures and Triumphs in Space* (New York: Random House, 1981), 54–55.

4. Or even a third, if one counts the *Lunik* flight around the moon. See Walter A. McDougall, . . . *the Heavens and the Earth: A Political History of the Space Age* (New York: Basic Books, 1985), 244.

5. See, for example, Charles Murray and Catherine Bly Cox, *Apollo: The Race to the Moon* (New York: Touchstone Books, 1989), 77; and Henry C. Dethloff, *Suddenly Tomorrow Came: A History of the Johnson Space Center* (Washington, DC: NASA, 1993), 29

6. *Pravda*, 12 May 1961, quoted in McDougall, . . . *the Heavens and the Earth*, 245.

7. Quoted in John M. Logsdon, *The Decision to Go to the Moon: Project Apollo and the National Interest* (Chicago: University of Chicago Press, 1970), 101.

8. Quoted in McDougall, . . . *the Heavens and the Earth*, 245.

9. Quoted in William B. Breuer, *Race to the Moon: America's Duel with the Soviets* (Westport, CT: Praeger, 1993), 163.

10. *Washington Post*, 13 April 1961, A18.

11. Quoted in Hugh Sidey, *John F. Kennedy, President* (New York: Atheneum, 1964), 99.

12. Quoted in U.S. Congress, House of Representatives, *Toward the Endless Frontier: History of the Committee on Science and Technology, 1959–1979* (Washington, DC: GPO, 1990), 82

13. *New York Times*, 17 April 1961, 5.

14. Quoted in Roger D. Launius, *NASA: A History of the U.S. Civil Space Program* (Malabar, FL: Krieger, 1994), 59.

15. *Congressional Record*, Senate, 8 May 1961, 7465.

16. Congressional Record, House, 24 May 1961, 8828.

17. Quoted in Lester A. Sobel, *Space: From Sputnik to Gemini* (New York: Facts on File, 1965), 120.

18. Quoted in T. A. Heppenheimer, *Countdown: A History of Spaceflight* (New York: Wiley, 1997), 192. In 1962, Pavel Popovich carried a famous (in the USSR, at any rate) photograph of Lenin at age four aboard *Vostok IV*, which orbited the earth 48 times. See Louis Fischer, *The Life of Lenin* (New York: Harper and Row, 1964), 1.

19. Oberg, *Red Star in Orbit*, 52.

20. McDougall, . . . *the Heavens and the Earth*, 247.

21. Reviewed in ibid., 246–247.

22. Of course, it could have been worse. *Vanguard*, after all, had blown up on the launch pad.

23. Launius, *NASA*, 59.

24. Quoted in U.S. Congress, House of Representatives, *Toward the Endless Frontier*, 82.

25. Quoted in Logsdon, *The Decision to Go to the Moon*, 103.

26. In some respects, of course, the general U.S. reaction to the flight was a bit milder than to that of *Sputnik I*. With the Space Age by now nearly four years old, many of the more irrational fears that grew out of the unfamiliarity of spaceflight had largely subsided. In addition, unlike 1957, the American public was aware that at least the country *had* a space program, and that it had been making steady, if slow, progress. Finally, *Vostok* had not been quite the surprise that *Sputnik* had been. A *Newsweek* story a month earlier, for example, had predicted the Soviet flight would come first (see Murray and Cox, *Apollo*, 76).

27. Of course, as the last chapter showed, this was not exactly correct. Many officials in the Eisenhower administration were aware that a Soviet satellite launch was imminent.

28. Shortly after *Sputnik*, for example, President Eisenhower stated, "There never has been one nickel asked for accelerating the [space] program. Never has it been considered as a race." See "The President's News Conference of October 9, 1957," *Public Papers of the Presidents: Dwight D. Eisenhower, 1957* (Washington, DC: GPO, 1958), 720.

29. For a brief period, it actually appeared as though the United States had a chance to win. NASA could have launched Shepard in March, but decided instead to conduct an additional test of the Redstone booster.

30. In addition, they had also had Webb as a witness on April 10 and George Low of NASA on April 11. See U.S. Congress, House of Representatives, *Toward the Endless Frontier*, 79.

31. Robert C. Seamans Jr., *Aiming at Targets: The Autobiography of Robert C. Seamans, Jr.* (Washington, DC: NASA SP-4106, 1996), 85.

32. U.S. Congress House of Representatives, *Toward the Endless Frontier*, 83

33. Quoted in Michael R. Beschloss, "Kennedy and the Decision to Go to the Moon," in *Spaceflight and the Myth of Presidential Leadership*, ed. Roger D. Launius and Howard E. McCurdy, 51–67, 61 (Urbana: University of Illinois, 1997).

34. This exchange is described in W. Henry Lambright, *Powering Apollo: James E. Webb of NASA*, Johns Hopkins New Series in NASA History (Baltimore: Johns Hopkins University Press, 1995), 93–94.

35. Tom Wolfe, *The Right Stuff* (New York: Bantam Books, 1979), 227–228.

36. Lambright, *Powering Apollo*, 94–95.

37. Lyndon Baines Johnson, *The Vantage Point: Perspectives on the Presidency, 1963–1969* (New York: Holt, Rinehart, and Winston, 1971), 280.

38. Arthur M. Schlesinger Jr., *A Thousand Days: John F. Kennedy in the White House* (Boston: Houghton Mifflin, 1965), 318.

39. Although there is certainly evidence that some congressmen, particularly Republican opponents of Apollo believed at the time that Kennedy was attempting to divert public attention with Apollo. See, for example, Representative

James D. Weaver's comments in *Congressional Record*, House, 28 October 1963, 20385.

40. Johnson biographer Robert Dallek has suggested—although on what basis is unclear—that Kennedy also wished to have the vice president in a position to take some of the criticism if the program suffered any sort of major setback. See Robert Dallek, "Johnson, Project Apollo, and the Politics of Space Program Planning," in *Spaceflight and the Myth of Presidential Leadership*, ed. Roger D. Launius and Howard E. McCurdy, 71–73 (Urbana: University of Illinois Press, 1997).

41. Ibid; see also Doris Kearns Goodwin, *Lyndon Johnson and the American Dream* (New York: Harper and Row, 1976), 162–164.

42. Beschloss, "Kennedy and the Decision to Go to the Moon," 58

43. Although there certainly were some, such as Science Advisor Jerome Wiesner and Budget Director David Bell, who were at least cautious about Apollo.

44. Quoted in Beschloss, "Kennedy and the Decision to Go to the Moon," 58

45. Memorandum from Robert S. McNamara for the Vice President, 21 April 1961, in LBJL and John M. Logsdon, ed., *Exploring the Unknown: Selected Documents in the History of the U.S. Civil Space Program*, vol. 1, *Organizing for Exploration* (Washington, DC: NASA, 1995), 424.

46. Beschloss, "Kennedy and the Decision to Go to the Moon," 57. See also McDougall, . . . *the Heavens and the Earth*, 322.

47. There is a very large body of literature examining President Kennedy's decision to send Americans to the moon. Indeed, it is probably one of the most heavily studied areas in all of space history. See Logsdon, *The Decision to Go to the Moon*, and "The Evolution of U.S. Space Policy and Plans," in *Exploring the Unknown: Selected Documents in the History of the U.S. Civil Space Program*, vol. 1, *Organizing for Exploration*, ed. John M. Logsdon 377–393, especially 379–381 (Washington, DC: NASA, 1995); Murray and Cox, *Apollo*, chap. 5; Launius, *NASA*, chap. 5; and Leonard Mandelbaum, "Apollo: How the United States Decided to Go to the Moon," *Science*, 14 February 1969, 649–654.

48. National Aeronautics and Space Administration, Office of Program Planning and Evaluation, *The Long Range Plan of the National Aeronautics and Space Administration* (Washington, DC: NASA, 16 December, 1959) in NASA Historical Reference Collection (hereafter NHRC). Excerpts from this report can be found in Logsdon, ed., *Exploring the Unknown*, 1:403–407.

49. In any event, there is no evidence that anyone in the Eisenhower administration (or, for that matter, Kennedy's prior to April) was particularly committed to this "goal."

50. John F. Kennedy, Memorandum for the Vice President, 20 April 1961, in John F. Kennedy Library (hereafter JFKL). A copy can also be found in Launius, *NASA*, 173–174, and in Logsdon, ed., *Exploring the Unknown*, 1:423–424; emphasis added.

51. "President's News Conference, 21 April 1961," *Public Papers of the Presidents: John F. Kennedy, 1961* (Washington, DC: GPO, 1962), 310 (hereafter *Public Papers, 1961*).

NOTES TO CHAPTER 5   205

52. Although he did note that, if it were to waive some safety requirements and limit the mission to a single cosmonaut, the USSR was capable of achieving a circumlunar flight by 1962 or 1963.

53. Letter from Wernher von Braun to the Vice President of the United States, 29 April 1961, in NHRC and Logsdon, ed., *Exploring the Unknown*, 1:429–433.

54. Memorandum from Robert S. McNamara for the Vice President (21 April 1961).

55. Lyndon B. Johnson, Memorandum for the President, 28 April 1961, in NHRC and Logsdon, ed., *Exploring the Unknown* 1:427–428; emphasis added.

56. Ibid.

57. Letter from James E. Webb and Robert S. McNamara to Lyndon B. Johnson, 8 May 1961, in NHRC and Logsdon, ed., *Exploring the Unknown*, 1:443–444; emphasis added.

58. Ibid., 444; emphasis added.

59. Logsdon, "The Evolution of U.S. Space Policy and Plans," 380.

60. "Special Message to Congress on Urgent National Needs, delivered in person before a Joint Session," 25 May 1961, *Public Papers, 1961*, 403.

61. Ibid., 405.

62. Ibid.

63. Quoted in John W. Finney, "President Links Space to Growth," *New York Times*, 1 February 1962, A1.

64. Although for years after, NASA would gamely point to the moon program's spin-offs.

65. "Expanded Effort Viewed as a Cold War Need, Survey Shows," *Aviation Week and Space Technology*, 5 February 1962, 34.

66. "President's News Conference," 17 July 1963, *Public Papers of the Presidents: John F. Kennedy, 1963* (Washington, DC: GPO, 1964), 568 (hereafter *Public Papers, 1963*).

67. "President's News Conference," 24 April 1963, *Public Papers, 1963*, 144.

68. "President's News Conference," 17 July 1963, *Public Papers, 1963*, 305.

69. "President's News Conference," 22 August 1963, *Public Papers, 1963*, 633.

70. Ibid. See also "President's News Conference," 3 April 1963, *Public Papers, 1963*, 310. For a general review of Kennedy's use of such imagery, see Linda T. Krug, *Presidential Perspectives on Space Exploration: Guiding Metaphors from Eisenhower to Bush* (New York: Praeger, 1991), chap. 2.

71. Remarks at the dedication of the Aerospace Medical Health Center, San Antonio, Texas, 21 November 1963 in *Public Papers, 1963*, 882–883.

72. Quoted in Beschloss, "Kennedy and the Decision to Go to the Moon," 58

73. See McDougall, . . . *the Heavens and the Earth*, 389–390; William David Compton, *Where No Man Has Gone Before: A History of Apollo Lunar Exploration Missions* (Washington, DC: NASA SP-4214, 1989), chap. 1.

74. That view in itself represented something of a reversal for Bush, since (as seen in chapter 3) he had urged Congress not to pursue the technologically infeasible ICBM.

75. G. Pascal Zachary, *Endless Frontier: Vannevar Bush, Engineer of the American Century* (New York: Free Press, 1997), 388–391.

76. Reported in James L. Kauffmann, *Selling Outer Space: Kennedy, the Media, and Funding for Project Apollo, 1961–1963* (Tuscaloosa: University of Alabama Press, 1994), 106.

77. Ibid., 290.

78. In fact, the friction between perceived political necessity and what was technologically possible was evident from the very beginning of the Apollo program. The original draft of the Urgent National Needs speech called for the lunar landing to take place by 1967, which, as the 50th anniversary of the Bolshevik Revolution, was generally assumed to be the year that the Soviets would make their attempt (assuming that they would make one at all). Reportedly, some members of the administration felt that it was not prudent to select so specific a date so near in the future (which, given that the United States up to that point had only managed to put a man in space for 15 minutes, was probably a wise precaution). Thus, the less precise "before this decade is out" language was substituted. See Logsdon, "The Evolution of U.S. Space Policy and Plans," 380–381.

79. Murray and Cox, *Apollo*, 109–110.

80. Roger E. Bilstein, *Stages to Saturn: A Technological History of the Apollo/Saturn Launch Vehicle*, NASA History Series (Washington, DC: NASA, 1980), 67.

81. See Director, Bureau of the Budget, Memorandum for the President, Draft, 13 November 1962, in Logsdon, ed., *Exploring the Unknown*, 1:456.

82. See Richard Lewis, "The Kennedy Effect," *Bulletin of the Atomic Scientists* 24 (March 1968): 2–5

83. *Atlanta Constitution*, 21 September 1963, 2–3; "Kennedy's Space Boomerang," *Congressional Record*, 10 October 1963, A6352–A6353; "Lunar Landing Lunacy," *Congressional Record*, 3 October 1963, 18614–18615.

84. Reported in *Congressional Record*, 24 May, 1963, 8203.

85. See his comments in *Congressional Record*, 19 November 1962, 22363; 19 November 1963, 22370; and 20 November 1963, 22451.

86. For an overview of congressional debate of the Apollo program, see Kauffmann, *Selling Outer Space*, chap. 6.

87. Director, Bureau of the Budget, Memorandum for the President, Draft, 13 November 1962, in Logsdon, eds., *Exploring the Unknown*, 1:456.

88. For a full account of the Lovell affair and the reactions to it, see Edward Clinton Ezell and Linda Neuman Ezell, *The Partnership: A History of the Apollo–Soyuz Test Project* (Washington, DC: NASA, 1978), 49–51; and Dodd L. Harvey and Linda C. Ciccoritti, *US–Soviet Cooperation in Space* (Miami: Center for Advanced International Studies, University of Miami, 1974), 112–121. The com-

plete text of the Lovell letter can be found in *Congressional Record*, 9 August 1963, 14701–14702.

89. Representative George P. Miller of California, for example, stated on the floor of the House that

> nothing would please a rival power which seeks leadership in the field of outer space than to have us accept their word that they are going to the moon. It is a fine ploy to lead us into a sense of security and then to turn around and destroy our prestige.

See *Congressional Record*, House, 29 October 1963, 20438.

90. "President's Press Conference," 17 July 1963, *Public Papers, 1963*, 567–568.

91. For a more complete discussion of these events, see W. D. Kay, "John F. Kennedy and the Two Faces of the U.S. Space Program, 1961–1963," *Presidential Studies Quarterly* 28.3 (1998): 573–586.

92. "Address before the 18th General Assembly of the United Nations," 20 September 1963, *Public Papers, 1963*, 695–696.

93. John M. Logsdon, "The Development of International Space Cooperation," in *Exploring the Unknown: Selected Documents in the History of the U.S. Civil Space Program*, vol. 2, *External Relationships*, ed. John M. Logsdon, 12–13 (Washington, DC: NASA, 1996).

94. Beschloss, "Kennedy and the Decision to Go to the Moon," 63.

95. Which still might not have been a simple matter. Noting the "heavy costs" of the lunar landing goal in his Urgent National Needs speech, Kennedy declared that "[i]f we were to go only halfway, or reduce our sights in the face of difficulty, it would be better not to go at all" (Special Message to Congress on Urgent National Needs, Delivered in Person before a Joint Session," 25 May 1961, *Public Papers, 1961*, 405). In other words, the president appears to have deliberately deprived himself of any sort of graceful exit strategy.

96. Although the 1967 target had been omitted from Kennedy's speech (see note 75), it was still the date guiding NASA planning. It would slip to 1968 due to budget cuts in the mid-1960s and to 1969 due to the *Apollo I* fire. See the discussion in chapter 6.

97. Director, Bureau of the Budget, Memorandum for the President, Draft, 13 November 1962, in Logsdon, ed., *Exploring the Unknown*, 1:457.

98. Ibid., 458; emphasis added.

99. James E. Webb, Administrator, NASA, to the President, 30 November 1962, in ibid., 461–467.

100. See note 50.

101. Director, Bureau of the Budget, Memorandum for the President, Draft, 13 November 1962, in Logsdon, ed., *Exploring the Unknown*, 1:457.

102. This quote and the following passage are from John F. Kennedy, Memorandum for the Vice President, 9 April 1963, in NHRC, JFKL, and Logsdon, ed., *Exploring the Unknown*, 1:467–468.

NOTES TO CHAPTER 5

103. See Logsdon's comments on the memo in "The Evolution of U.S. Space Policy and Plans," 467.

104. John F. Kennedy, Memorandum for the Vice Pesident, 9 April 1963, in NHRC, JFKL, and Logsdon, ed., *Exploring the Unknown*, 1:468.

105. This passage and the quotes in the preceding paragraph are from Lyndon B. Johnson, vice president, to the President, 13 May 1963, with attached report, in Logsdon, ed., *Exploring the Unknown*, 1:468–473.

106. *Congressional Record*, 10 October 1963, 19240.

107. Quoted in ibid., 128. See also U.S. Congress, House of Representatives, *Toward the Endless Frontier*, 175.

108. *Congressional Record*, 10 October 1963, 19240.

109. For the complete text of Kennedy's letter, see National Aeronautics and Space Administration, *Astronautics and Aeronautics, 1963* (Washington, DC: NASA, 1964), 349–350.

110. National Security Action Memorandum No. 271 to the Administrator, National Aeronautics and Space Administration, "Cooperation with the USSR on Outer Space Matters," 12 November 1963, in NHRC.

111. "The President's News Conference of October 9, 1963," *Public Papers, 1963*, 769.

112. See note 71.

113. Actually, there are a number of interpretations. See Kay, "John F. Kennedy and the Two Faces of the U.S. Space Program."

114. This would explain, for example, the president's coolness to the Lovell letter, which was made public just a couple of months before his UN speech. Simply accepting an offer made by the Soviets (assuming that it was genuine), would not have had the same effect. Under the strategy outlined here, it would have been essential for the United States to be seen as making the offer first.

115. See *New York Times*, 21 September 1963, A9. Of course, such an integration did prove possible 10 years later during the *Apollo–Soyuz* Test Project (see next chapter).

116. Quoted in a memo from E. C. Welsh to Andrew Hatcher, 1 November 1963, in JFKL. Actually, if this account of Kennedy's intent is correct (and if Khrushchev recognized what was behind the proposal), then declining without refusing outright was perhaps his only available response.

117. *Congressional Record*, 10 October 1963, 19260. It should be noted, however, that some members supported the idea. See remarks by Senator J. William Fulbright in *Congressional Record*, 2 October 1963, 18543–18544. In addition, Congressman George Brown of California, then serving his first term, later recalled favoring it. Interview with George Brown, 10 July 1997; also interview with Don Fuqua, 10 July 1997.

118. *Congressional Record*, 19 November 1963, 22357.

119. *US–USSR Cooperation in Space Research Programs*, Report to the President from the Administrator, NASA, 31 January, 1964, in NHRC.

120. Quoted in Dallek, "Johnson, Project Apollo, and the Politics of Space Program Planning," 75.

121. U.S. National Aeronautics and Space Administration, *NASA Pocket Statistics, 1996* (Washington, DC: NASA, 1996), C–28.

122. Launius, *NASA*, 70.

123. Space did not play much of a role in the 1964 campaign. Senator Barry Goldwater, the GOP candidate, was one of several Republican officials who had called for more money for military space programs, and had criticized Project Apollo on the grounds that it was irrelevant to national security. Thus, while, in view of public opinion at the time, it may have been, as Dallek asserts, "superb politics," there is no evidence, that Johnson's or Goldwater's position on Apollo actually swayed many voters one way or the other. See Dallek, "Johnson, Project Apollo, and the Politics of Space Program Planning," 78–79.

124. Of course the previous year it had been zero: the only attempt, the ill-fated *Vanguard*, blew up in December 1958.

125. U.S. National Aeronautics and Space Administration, *Aeronautics and Space Report of the President* (Washington, DC: GPO, ____), app. A-1. It should be noted that these figures include *all* U.S. spacecraft, not just those launched by NASA.

126. Ibid., app. C.

127. James Oberg has written widely about his belief that the USSR did indeed intend to beat the United States to the moon, but could not complete its large booster rocket in time. See his *Red Star in Orbit*, chap. 7.

128. Quoted in ibid., 77.

129. U.S. National Aeronautics and Space Administration, *Aeronautics and Space Report of the President, Fiscal Year 1964 Activities* (Washington, DC: GPO, 1965), 120.

130. Launius, *NASA*, chap. 8.

Chapter 6

1. Recounted in Charles Murray and Catherine Bly Cox, *Apollo: The Race to the Moon* (New York: Touchstone Books, 1989), 370.

2. Linda T. Krug, *Presidential Perspectives on Space Exploration: Guiding Metaphors from Eisenhower to Bush* (New York: Praeger, 1991), 46.

3. Richard Hutton, *The Cosmic Chase* (New York: Mentor, 1981), 104.

4. U.S. Congress, House of Representatives, *United States Civilian Space Programs 1957–1978: Report Prepared for the Subcommittee on Space Science and Applications of the Committee on Science and Technology*, 97th Cong., 1st sess., January 1981 (Washington, DC: GPO, 1981), 380.

5. U.S. National Aeronautics and Space Administration, *Aeronautics and Space Report of the President, Fiscal Year 1994 Activities* (Washington, DC: NASA, 1995), app. E-1.

6. There is one somewhat similar, albeit smaller in scale, policy issue.

7. Memo for the President from Charles L. Schultze (BOB director) with attached letter to James E. Webb, 4 October 1966, copy in the Lyndon B. Johnson Presidential Library (hereafter LBJL).

8. Letter to the President from James E. Webb, 16 May 1966, in LBJL.

9. Quoted in John M. Logsdon, "The Evolution of U.S. Space Policy and Plans," in *Exploring the Unknown: Selected Documents in the History of the U.S. Civil Space Program*, vol. 1, *Organizing for Exploration*, ed. John M. Logsdon, 382 (Washington, DC: NASA, 1995).

10. Ibid. See also W. Henry Lambright, *Powering Apollo: James E. Webb of NASA*, Johns Hopkins New Series in NASA History (Baltimore: Johns Hopkins University Press, 1995) and "James E. Webb: A Dominant Force in 20th Century Public Administration," *Public Administration Review* 53.2 (1993): 95–99.

11. U.S. National Aeronautics and Space Administration, Future Programs Task Group, "*Summary Report, January 1965,*" in John M. Logsdon, ed., *Exploring the Unknown: Selected Documents in the History of the U.S. Civil Space Program*, vol. 1, *Organizing for Exploration* (Washington, DC: NASA, 1995), 473–489.

12. Letter to the President from James E. Webb, 16 February 1965, in LBJL.

13. U.S. Congress, House of Representatives, *1966 NASA Authorization: Hearings before the Subcommittee on Manned Space Flight, Committee on Science and Astronautics*, 89th Cong., 1st sess. (3, 4, 11, 16, and 17 March 1965), 134–135.

14. Letter to the President from James E. Webb, 26 August 1966, in LBJL and in Logsdon, ed., *Exploring the Unknown*, 1:491.

15. This program, in a greatly reduced form, was renamed Skylab in 1970.

16. Lambright, *Powering Apollo*, 139.

17. Memorandum to Associate Administrator for Manned Space Flight from James E. Webb, "Termination of the Contract for Procurement of Long Lead Time Items for Vehicles 516 and 517," 1 August 1968, in Logsdon, ed., *Exploring the Unknown*, 1:494–495.

18. This is not to suggest that these events are necessarily related. Indeed, there is no little mystery surrounding Webb's departure from NASA. See Lambright, *Powering Apollo*, 200–205.

19. Quoted in U.S. Congress, House of Representatives, *Toward the Endless Frontier: History of the Committee on Science and Technology, 1959–1979* (Washington, DC: GPO, 1990), 190.

20. Ibid., 191.

21. John Noble Wilford, "U.S. Space Effort Is Shrinking after Era of Growth," *New York Times*, 16 April 1968, 1–2.

22. Robert Dallek, "Johnson, Project Apollo, and the Politics of Space Program Planning," in *Spaceflight and the Myth of Presidential Leadership*, ed. Roger D. Launius and Howard E. McCurdy, 80 (Urbana: University of Illinois, 1997). See also Lyndon Baines Johnson, *The Vantage Point: Perspectives on the Presidency, 1963–1969* (New York: Holt, Rinehart, and Winston, 1971).

23. Quoted in Dallek, "Johnson, Project Apollo, and the Politics of Space Planning."

24. Quoted in Michael R. Beschloss, "Kennedy and the Decision to Go to the Moon," in *Spaceflight and the Myth of Presidential Leadership*, ed. Roger D. Launius and Howard E. McCurdy (Urbana: University of Illinois, 1997), 58.

25. Letter to the President from James E. Webb, 26 August 1966, in Logsdon, ed., *Exploring the Unknown*, 1:492.

26. Letter to the President from James E. Webb, 10 July 1967, in LBJL.

27. Signing Statement on S. 1296, the FY 1968 NASA Authorization Bill, in LBJL.

28. Confidential memorandum to James E. Webb from the President, 29 September 1967, in LBJL.

29. Quoted in Dallek, "Johnson, Project Apollo, and the Politics of Space Program Planning," 84.

30. For a general view of President Johnson's rhetoric on space, see Krug, *Presidential Perspectives on Space Exploration*, chap. 2.

31. This on top of the Kennedy administration's initiatives, which had already pushed space R&D collaboration to unprecedented levels.

32. "Telephone Conversation between the President and Astronauts McDivitt and White, 7 June 1965," *Public Papers of the Presidents: Lyndon Johnson, vol. 2, 1965* (Washington, DC: GPO, 1966), 646 (hereafter *Public Papers, 1965*).

33. "Remarks in Houston at the Manned Spacecraft Center, 11 June 1965," *Public Papers, 1965*, 2:656.

34. "Remarks at a Ceremony in the State Department Auditorium in Honor of the Gemini 4 Astronauts, 17 June 1965," *Public Papers, 1965*, 2:685.

35. "Remarks upon Accepting the Robert H. Goddard Trophy, 16 March 1966," *Public Papers of the Presidents: Lyndon Johnson, 1966*, vol. 1 (Washington, DC: GPO, 1967), 330.

36. "News Conference, 29 August 1965," *Public Papers, 1965*, 2:945.

37. See Confidential Memorandum to Secretary Dean Rusk regarding President Johnson on UN Affairs, 26 November 1963, in LBJL. The substance of the discussion is also recounted in Theodore H. White, *The Making of the President, 1964* (New York: Signet, 1965), 63–64.

38. "Outer Space Treaty Chronology," in LBJL.

39. U.S. Department of State, *Space Goals after the Lunar Landing*, October 1966, in LBJL.

40. See Dallek, "Johnson, Project Apollo, and the Politics of Space Program Planning," 82.

41. See, for example, Memorandum for the President from Charles L. Schultze, 1 September 1966 and 16 December 1966, both in LBJL.

42. Memorandum for the President from Charles L. Schultze, 11 August 1967, in LBJL.

43. Cited in Dallek, "Johnson, Project Apollo, and the Politics of Space Program Planning," 83.

44. Herbert E. Krugman, "Public Attitudes toward the Apollo Space Program, 1965–1975," *Journal of Communication* (Autumn 1977): 87–93.

45. Thomas Hughes, *American Genesis: A Century of Invention and Technological Enthusiasm, 1870–1970* (New York: Viking, 1989), chap. 9; Bruce L. R. Smith, *American Science Policy since World War II* (Washington, DC: Brookings Institution, 1990), chap. 4; Daniel J. Kevles, *The Physicists: The History of a Scientific Community in Modern America* (Cambridge, MA: Harvard

University Press, 1995), especially chaps. 24 and 25; Roger D. Launius, *NASA: A History of the U.S. Civil Space Program* (Malabar, FL: Krieger, 1994), 94–95; and John McDermott, "Technology: The Opiate of the Intellectuals," in *Technology and the Future*, 7th ed., ed. Albert H. Teich, 77–92 (New York: St. Martin's, 1997).

46. See (among many others) W. D. Kay, *Can Democracies Fly in Space? The Challenge of Revitalizing the U.S. Space Program* (Westport, CT: Praeger, 1995), 55; Logsdon, "The Evolution of U.S. Space Policy and Plans," 383; Dallek, "Johnson, Project Apollo, and the Politics of Space Program Planning," 81; and Lambright, *Powering Apollo*, 193.

47. See, for example, Memorandum to the President from Charles L. Schultze, 20 September 1966, in LBJL.

48. Letter to the President from James E. Webb, 26 August 1966, in LBJL and Logsdon, ed., *Exploring the Unknown*, 1:491.

49. U.S. Congress, House of Representatives, *Toward the Endless Frontier*, 191.

50. Ibid., 193.

51. Reported in Lambright, *Powering Apollo*, 193.

52. Reported in Jim F. Heath, *Decade of Disillusionment: The Kennedy–Johnson Years* (Bloomington: Indiana University Press, 1975), 264.

53. Cited in Dallek, "Johnson, Project Apollo, and the Politics of Space Program Planning," 83

54. Quoted in ibid.

55. Rip Bulkeley, *The Sputniks Crisis and Early United States Space Policy* (Bloomington: Indiana University Press, 1990).

56. Arnold S. Levine, *Managing NASA in the Apollo Era* (Washington, DC: NASA, 1982), chap. 9.

57. See U.S. Congress, House of Representatives, *Toward the Endless Frontier*, 195–205; Lambright, *Powering Apollo*, chap. 8.

58. Quoted in Dallek, "Johnson, Project Apollo, and the Politics of Space Program Planning," 84.

59. U.S. National Science Foundation, National Science Board, *Science and Engineering Indicators, 1996* (NSB-96-21) (Washington, DC: NSF, 1996), app. 4, table 4-3. See also Alexander J. Morin, *Science and Politics* (Englewood Cliffs, NJ: Prentice-Hall, 1993), chap. 4

60. Quoted in Hutton, *The Cosmic Chase*, 80.

61. This does not, however, seem to represent any sort of acknowledgment of a mistake.

62. Even programs with legitimate scientific research aims or practical applications, such as communication and weather satellites, also served American Cold War interests: as has been noted, officials frequently pointed out that the United States "led" the Soviet Union in these areas.

63. "President's News Conference, 17 July 1963," *Public Papers, 1963*, 568.

64. James Oberg, *Red Star in Orbit: The Inside Story of Soviet Failures and Triumphs in Space* (New York: Random House, 1981), 90.

65. Moreover, although this had not been known in the West at the time, the flight did not even come close to achieving its objectives. Apparently the plan had been to eclipse the U.S. Gemini program by having two spacecraft dock and exchange crews (a mission that was accomplished two years later). The second Soyuz craft, however, was never launched. See ibid., 94.

66. U.S. Congress, Office of Technology Assessment, *Exploring the Moon and Mars: Choices for the Nation* (Washington, DC: OTA, 1991), 50–53.

67. Hutton, *The Cosmic Chase*, 81.

68. See Roger E. Bilstein, *Stages to Saturn: A Technological History of the Apollo/Saturn Launch Vehicle*, NASA History Series (Washington, DC: NASA, 1980).

69. "Remarks Made at the NASA Manned Spacecraft Center," 11 June 1965, *Public Papers, 1965*, 2:656.

70. "Remarks upon Accepting the Robert H. Goddard Trophy." In fact, on the day of his assassination in 1963, President Kennedy had planned to say that "there was no longer any fear that a Communist lead in space would become a basis for military superiority" (quoted in Dallek, "Johnson, Project Apollo, and the Politics of Space Program Planning," 75).

71. Memorandum for the Vice President from E. C. Welsh, 17 November 1966, in LBJL.

72. On two occasions, Soviet Vostok spacecraft flew relatively close to one another, but this was due exclusively to the precision of the orbital paths on which they were launched initially. In neither case could the pilot of either craft maneuver closer to the other on their own.

73. Memorandum for the President from Charles L. Schultze, 1 September 1966, in LBJL, 4–5; emphasis in the original.

74. "Statement by the President on the Tenth Anniversary of NASA," in LBJL.

75. "Nigerians in Lagos See U.S. Taking Space Lead from Soviets," USIA Report M-402-65 (USIA-IRS/AA), 30 September 1965, in LBJL.

76. Memorandum for W. W. Rostow from Robert L. Sansom, 24 December 1968, in LBJL.

77. Quoted in Dallek, "Johnson, Project Apollo, and the Politics of Space Program Planning," 75.

78. U.S. Congress, House of Representatives, *Toward the Endless Frontier*, 171–172; Walter A. McDougall, *...the Heavens and the Earth: A Political History of the Space Age* (New York: Basic Books, 1985), 391–393.

79. Memorandum for the President from James E. Webb, 1 April 1966, in LBJL.

80. Memorandum for the President from James E. Webb, 16 May 1966, in LBJL.

81. Quoted in Letter to the President from James E. Webb, 26 August 1966, in LBJL and Logsdon, ed., *Exploring the Unknown*, 1:491.

82. Quoted in letter to the President from James E. Webb, 5 October 1968, in LBJL, 3–4.

83. Memorandum for Charles Schultze from James E. Webb, 6 November 1967, in LBJL.

84. Quoted in letter to the President from James E. Webb, 5 October 1968, in LBJL, 3. See also T. A. Heppenheimer, *Countdown: A History of Spaceflight* (New York: Wiley, 1997), 238.

85. Recounted in Dallek, "Johnson, Project Apollo, and the Politics of Space Program Planning," 86.

86. Quoted in Letter to the President from James E. Webb, 5 October 1968, in LBJL, 4.

87. Ibid., 5.

88. Dallek, "Johnson, Project Apollo, and the Politics of Space Program Planning," 80.

89. Oberg, *Red Star in Orbit*, chap. 7.

90. Not the least of which was Vietnam.

91. U.S. National Aeronautics and Space Administration, *Aeronautics and Space Report of the President*.

92. National Academy of Sciences, National Research Council, Physics Survey Committee, *Physics in Perspective*, vol. 1 (Washington, DC: National Academy Press, 1972), 2.

93. Cited in Kevles, *The Physicists*, 416.

94. An electron volt is the amount of energy of motion acquired by an electron by an electrical potential of one volt. The first postwar accelerators generally were in the range of 4–5 billion electron volts (abbreviated GeV). By contrast, the Superconducting Super Collider, which was canceled in 1993, was to operate in the neighborhood of *40 trillion* electron volts. See W. D. Kay, "Particle Physics and the Superconducting Super Collider: The Collaboration That Wasn't," in *International Collaboration in Science and Technology among Advanced Industrial Societies*, ed. Vicki L. Golich and W. D. Kay (London: Routledge, forthcoming).

95. Quoted in Kevles, *The Physicists*, 416.

96. Quoted in Daniel S. Greenberg, *The Politics of Pure Science* (New York: New American Library, 1967), 260.

97. See, for example, John Fisher, "Why Our Scientists Are About to Be Dragged, Moaning, into Politics," *Harper's* (September 1966): 27.

98. Quoted in Kevles, *The Physicists*, 417–418.

99. That was not the end of the matter, however. A few years later, the project was revived (albeit not without controversy), and a large accelerator was built at a site near Chicago (later named the Fermi National Laboratory, or Fermilab).

100. Kevles, *The Physicists*, 421.

101. Quoted in ibid.

102. For a review of these arguments, see Kay, *Can Democracies Fly in Space?* 19–20.

103. Laird was represented by Air Force Secretary (and former NASA Associate Administrator) Robert Seamans.

104. Memorandum from the President to the Vice President, the Secretary of Defense, the Acting Administrator of the National Aeronautics and Space Administration, and the Science Advisor, 13 February 1969, in NASA History Office and in Logsdon, ed., *Exploring the Unknown*, 1:512–513.

105. Quoted in Logsdon, "The Evolution of U.S. Space Policy and Plans," 384.

106. Quoted in *New York Times*, 17 July 1969, 1.

107. Logsdon, "The Evolution of U.S. Space Policy and Plans," 384–385. See also Robert C. Seamans Jr., *Aiming at Targets: The Autobiography of Robert C. Seamans, Jr.* (Washington, DC: NASA SP-4106, 1996), 242–244.

108. See U.S. National Aeronautics and Space Administration, *America's Next Decade in Space: A Report for the Space Task Group* (Washington, DC: NASA, 1969).

109. Memorandum from Robert Mayo to the President, "Space Task Group Report," 25 September 1970, in Logsdon, ed., *Exploring the Unknown*, 1:544–546.

110. "Statement about the Future of the United States Space Program," 7 March 1970, *Public Papers of the Presidents: Richard M. Nixon, 1970* (Washington, DC: GPO, 1970), 250–253.

111. Although it should be noted that during his brief tenure in office, Gerald Ford made few definitive statements about the program. See, for example, Krug, *Presidential Perspectives on Space Exploration*, 56–59.

112. "Statement about the Future of the United States Space Program."

113. Memorandum from Frank Press to the President, 30 August 1977, Jimmy Carter Presidential Library (hereafter JCL).

114. Memorandum from Frank Press to Administrator, National Aeronautics and Space Administration; the Secretaries of State, Defense, Commerce, Agriculture, Interior, Energy; Director, Central Intelligence Agency; Assistant to the President for National Security Affairs; Director, Domestic Policy Staff; Director, Office of Management and Budget, "Civil Space and Aeronautics Policy," 8 May 1978, in JCL.

115. Presidential Directive NSC-37, "National Space Policy," 11 May 1978, in JCL. See also Logsdon, "The Evolution of U.S. Space Policy and Plans," 388–389.

116. White House Fact Sheet: "U.S. Civil Space Policy," 11 October 1978, in JCL.

117. Memorandum from Frank Press to the President, "Basic Research," 17 October 1978, in JCL.

118. U.S. Congress, Senate, *NASA Authorization for Fiscal Year 1970: Hearings before the Committee on Aeronautical and Space Sciences*, 91st Cong., 1st sess., Part 2 (1, 6, and 9 May 1969), 876.

119. U.S. Congress, Senate, *Nomination of Dr. Thomas O. Paine to Be Administrator of the National Aeronautics and Space Administration* (14 March 1969) (Washington, DC: GPO, 1969), 12

120. "Statement about the Future of the United States Space Program."

121. Memorandum from Frank Press; "Civil Space and Aeronautics Policy," in JCL, 14.

122. Presidential Directive NSC-37, "National Space Policy."

123. White House Fact Sheet: "U.S. Civil Space Policy."

124. "Statement about the Future of the United States Space Program."

125. "Statement Announcing Decision to Proceed with Development of the Space Shuttle," 5 January 1972, *Public Papers of the Presidents: Richard M. Nixon, 1972* (Washington, DC: GPO, 1973), 20–21.

126. Memorandum from Frank Press to the President, "Basic Research."

127. Presidential Directive NSC-37, "National Space Policy."

128. "Remarks and a Q&A Session with Editors and News Directors," 13 October 1978, *Public Papers of the Presidents: James E. Carter, 1978*, vol. 2 (Washington, DC: GPO, 1979), 1777. In marking the 10th anniversary of the *Apollo 11* landing, Carter noted that the primary benefit of the space program was to "help us manage our earth." See "Apollo 11 Anniversary" in JCL.

129. Louis Harris, *The Harris Survey Yearbook of Public Opinion, 1970* (New York: _____, 1971), 83–84.

130. Memorandum from Peter M. Flannigan to the President, 6 December 1969, in Logsdon, ed., *Exploring the Unknown*, 1:546.

131. Memorandum from James E. Webb to the President, 30 November 1962, in ibid., 466.

132. See, for example, National Aeronautics and Space Council, "U.S. Policy on Outer Space, 26 January 1960," in Logsdon, ed., *Exploring the Unknown*, 1:365.

133. "President's News Conference," 17 July 1963, *Public Papers of the President: John F. Kennedy, 1963* (Washington, DC: GPO, 1964), 568.

134. It should also be kept in mind that human spaceflight represents what is sometimes called a "lumpy" requirement, in that very large costs are required for it to be done at all, even at a relatively reduced level.

135. Memorandum from Caspar W. Weinberger, via George Schultz, to the President, "Future of NASA," 12 August 1971, in Logsdon, ed., *Exploring the Unknown*, 1:546–548. On one copy of the memo, Nixon wrote, "I agree with Cap." Although Weinberger does not explicitly mention human spaceflight, he was moved to write the memo by proposals, mostly within OMB, to reject the proposed space shuttle and to cancel *Apollo 16* and *Apollo 17*.

136. See John M. Logsdon, "The Space Shuttle Program: A Policy Failure?" *Science*, 30 May 1986, 1099–1105. For a view on Nixon's political views concerning space, see Stephen E. Ambrose, *Nixon: The Triumph of a Politician, 1962–1972* (New York: Simon and Schuster 1989), 498.

137. Interview with John Erlichman by John M. Logsdon, 6 May 1983, in NASA History Office. See also Timothy Ferris, "Some Like it Hot," *New York Review of Books* 44.14 (1997): 16–20.

138. Next to Apollo, NASA's efforts to secure approval for the shuttle is one of the most closely examined events in the history of the space program. See, for example, John M. Logsdon, "The Space Shuttle Decision: Technology and Political Choice," *Journal of Contemporary Business* 7 (1978): 13–30, "The Decision to Develop the Space Shuttle," *Space Policy* 2 (1986): 103–119, and "The Space Shuttle Program: A Policy Failure?" *Science*, 30 May 1986, 1099–1105; Thomas H. Johnson, "The Natural History of the Space Shuttle," *Technology in Society* 10 (1988): 417–424; Jeffrey S. Banks, "The Space Shuttle," in *The Technology Pork Barrel*, ed. Linda R. Cohen and Roger G. Noll, 179–216 (Washington, DC:

Brookings Institution, 1991); Dennis R. Jenkins, *The History and Development of the Space Transportation System: The Beginning through STS-50* (Melbourne Beach, FL: Broadfield, 1992; and Launius, *NASA*, chap. 9, and "NASA and the Decision to Build the Space Shuttle, 1969–1972," unpublished paper.

139. "Budget for Approved Shuttle $2.3 Billion Less Than NASA Wanted," *Space Business Daily* 60.4 (1972): 1–2.

140. Claude E. Barfield, "Technology Report/NASA Report Broadens Defense of Space Shuttle to Counter Critics' Attacks," *National Journal* 4 (1972): 1323–1332; emphasis added. See also Launius, "NASA and the Decision to Build the Space Shuttle, 1969–1972."

141. W. D. Kay, "Democracy and Super Technologies: The Politics of the Space Shuttle and Space Station *Freedom*," *Science, Technology, and Human Values* 19.2 (1994): 131–151.

142. See note 136.

143. For more on the problems of space science during this period, see (among many others) Bruce Murray, *Journey Into Space: The First Thirty Years of Space Exploration* (New York: Norton, 1989); John M. Logsdon, "Missing Halley's Comet: The Politics of Big Science," *Isis* 80 (1989): 254–280; and W. D. Kay, "Where No Nation Has Gone Before: Domestic Politics and the First International Space Science Mission," *Journal of Policy History* 5.3 (1993): 435–452.

144. Nick MacNeil, Carter–Mondale Transition Planning Group, to Stuart Eisenstat, Al Stern, David Rubenstein, Barry Blechman, and Dick Steadman, "NASA Recommendations, 31 January 1977," in Logsdon, ed., *Exploring the Unknown*, 1:559–574.

144. A May 23, 1978, memorandum from Stuart Eisenstat to Al Stern (in JCL) warns about the "distinct possibility of Soviet space spectaculars," but suggests that "we need not programmatically anticipate" them.

Chapter 7

1. Dilys M. Hill and Phil Williams, "The Reagan Presidency: Style and Substance" in *The Reagan Presidency: An Incomplete Revolution?* ed. Dilys M. Hill, Raymond A. Moore, and Phil Williams, 3–25 (New York: St. Martin's, 1990).

2. Graham K. Wilson, *Business and Politics: A Comparative Introduction*, 2nd ed. (Chatham, NJ: Chatham House, 1990), 53.

3. See, for example, Gillian Peele, "The Agenda of the New Right," in *The Reagan Presidency: An Incomplete Revolution?* ed. Dilys M. Hill, Raymond A. Moore, and Phil Williams, 29–47 (New York: St. Martin's, 1990).

4. Paul J. Quirk, "The Economy: Economics, Electoral Politics, and Reagan Economics," in *The Election of 1984*, ed. Michael Nelson, 155–187 (Washington, DC: Congressional Quarterly Press, 1985).

5. Which, incidentally, is a common feature of American political parties.

6. Benjamin Ginsberg and Martin Shefter, "A Critical Realignment? The New Politics, the Reconstituted Right, and the Election of 1984," in *The Election of 1984*, ed. Michael Nelson, 1–25 (Washington, DC: Congressional Quarterly Press, 1985).

7. Which may mean, in fact, that (the admonition at the beginning of the chapter notwithstanding) Reagan's may well have been the most pro-business administration since the 1920s. See Allen M. Kaufman, Marvin J. Karson, and Jeffrey Sohl, "Business Fragmentation and Solidarity: An Analysis of PAC Donations in the 1980 and 1982 Elections," in *Business Strategy and Public Policy*, ed. Alfred A. Marcus, Allen M. Kaufman, and David R. Beam, 119–136 (New York: Quorum, 1987); and Theodore J. Eismer and Phillip H. Pollock III, "The Retreat from Partisanship: Why the Dog Didn't Bark in the 1984 Election" in *Business Strategy and Public Policy*, 137–147.

8. Ginsberg and Shefter, "A Critical Realignment?" See also Theodore J. Lowi, "An Aligning Election: A Presidential Plebiscite," in *The Election of 1984*, ed. Michael Nelson, 277–301 (Washington, DC: Congressional Quarterly Press, 1985).

9. Irene S. Rubin, *The Politics of the Budgetary Process: Getting and Spending, Borrowing, and Balancing*, 2nd ed. (Chatham, NJ: Chatham House, 1993), 12.

10. Elizabeth Drew, "A Political Journal," *New Yorker*, 20 February 1984, 132.

11. Peter Goldman and Tony Fuller, *The Quest for the Presidency 1984* (New York: Bantam Books, 1985), 26.

12. Ibid., 27.

13. Hill and Williams, "The Reagan Presidency."

14. Lyn Ragsdale, "Politics, Not Science: The U.S. Space Program in the Reagan and Bush Years," in *Spaceflight and the Myth of Presidential Leadership*, ed. Roger D. Launius and Howard E. McCurdy, 133–171, 134, (Urbana: University of Illinois, 1997).

15. "Inaugural Address, 20 January 1981." *Public Papers of the Presidents: Ronald Reagan, 1981* (Washington, DC: GPO, 1982), 10.

16. Quoted in Charles T. Goodsell, *The Case for Bureaucracy: A Public Administration Polemic*, 3rd ed. (Chatham, NJ: Chatham House, 1985), 166.

17. Dilys M. Hill, "Domestic Policy in an Era of 'Negative' Government," in *The Reagan Presidency: An Incomplete Revolution?* ed. Dilys M. Hill, Raymond A. Moore, and Phil Williams, 161–178 (New York: St. Martin's, 1990).

18. Murray I. Wiedenbaum, "Government Power and Economic Performance," in *The United States in the 1980s*, ed. Peter Duigan and Alvin Rabushka, 181 (Reading, MA: Addison-Wesley, 1980).

19. Ibid., 175–176. See also Gillian Peele, "The Agenda of the New Right," *The Reagan Presidency: An Incomplete Revolution?* ed. Dilys M. Hill, Raymond A. Moore, and Phil Williams, 29–47 (New York: St. Martin's, 1990).

20. E. S. Savas, *Privatization: The Key to Better Government* (Chatham, NJ: Chatham House, 1987), 3.

21. U.S. President's Commission on Privatization, *Privatization: Toward More Effective Government* (Urbana: University of Illinois Press, 1988), 1.

22. Savas, *Privatization*, 288–291. See also James Q. Wilson, "The Rediscovery of Character: Private Virtue and Public Policy," *The Essential Neo-Conservative Reader*, ed. Mark Gerson, 291–304 (Reading, MA: Addison-Wesley,

1996); and Arthur A. Goldsmith, *Business, Government, Society: The Global Political Economy* (Chicago: Irwin, 1996), 26–28.

23. This does not include the use of voucher systems, which proponents advocate for services like education or housing (see E. S. Savas, *Privatizing the Public Sector: How to Shrink Government* [Chatham, NJ: Chatham House, 1982], especially 67–69), nor does it include deregulation, which some consider a form of privatization (see U.S. President's Commission on Privatization, *Privatization*, 2).

24. Savas, *Privatizing the Public Sector* and *Privatization*.

25. Bruce L. R. Smith, *American Science Policy since World War II* (Washington, DC: Brookings Institution, 1990), 125.

26. Savas, *Privatization*; U.S. President's Commission on Privatization, *Privatization*, 3–4.

27. See Richard D. Bingham, *Industrial Policy American Style: From Hamilton to HDTV* (Armonk, NY: Sharpe, 1998). Bingham points out in chapter 1 that, although the term *industrial policy* is fairly recent, debate over the policies themselves goes back to the time of Alexander Hamilton.

28. Nicolas Spulber, *The American Economy: The Struggle for Supremacy in the 21st Century* (New York: Cambridge University Press, 1995), 120.

29. David A. Stockman, *The Triumph of Politics: Why the Reagan Revolution Failed* (New York: Harper and Row, 1986), 157.

30. Goldsmith, *Business, Government, Society*, 366. See also Kevin P. Phillips, "U.S. Industrial Policy: Inevitable and Ineffective," *Harvard Business Review* (July/August 1992): 104–112.

31. W. W. Kaufman, "A Defense Agenda for Fiscal Years 1990–94," in *Restructuring American Foreign Policy*, ed. John Steinbruner, 4–77, 57 (Washington, DC: Brookings Institution, 1987). See also Phil Williams, "The Reagan Administration and Defence Policy," in *The Reagan Presidency: An Incomplete Revolution?* ed. Dilys M. Hill, Raymond A. Moore, and Phil Williams, 199–230 (New York: St. Martin's, 1990) .

32. Lou Cannon, *President Reagan: The Role of a Lifetime* (New York: Simon and Schuster, 1991), 92–93.

33. Quoted in ibid., 281–282.

34. Raymond A. Moore, "The Reagan Presidency and Foreign Policy," in *The Reagan Presidency: An Incomplete Revolution?* ed. Dilys M. Hill, Raymond A. Moore, and Phil Williams, 178–198, 183 (New York: St. Martin's, 1990).

35. Quoted in Cannon, *President Reagan*, 296

36. Ibid.

37. See, for example, David Dickson, *The New Politics of Science* (New York: Pantheon, 1984), especially chap. 3.

38. Quoted in Stockman, *The Triumph of Politics*, 151.

39. "Science and Technology for America's Future, 11 March 1986," Ronald Reagan Presidential Library (hereafter RRL).

40. Smith, *American Science Policy since World War II*, 129.

41. See U.S. National Science Foundation, National Science Board, *Science and Engineering Indicators, 1996* (NSB-96-21) (Washington, DC: NSF, 1996), 4–9.

42. U.S. Office of Science and Technology Policy, *Annual Science and Technology Report to the Congress, 1982* (Washington, DC: OSTP, 1982).
43. "Science and Technology for America's Future," in RRL.
44. Ibid.
45. U.S. National Science Foundation, *Science and Engineering Indicators, 1996*, app. 4, table 4-5. Figures are in current dollars. Some estimates place the increase at more than 100 percent. See Smith, *American Science Policy since World War II*, 133.
46. Quoted in Dickson, *The New Politics of Science*, 11–12.
47. U.S. National Science Foundation, *Science and Engineering Indicators, 1996*, 4–9.
48. Ibid.
49. See Sylvia Ostrey, *Governments and Corporations in a Shrinking World* (New York: Council on Foreign Relations, 1990). For a specific example, see Stockman, *The Triumph of Politics*, 114–115. For an overview of both sides of the argument, see Ernest Sternberg, *Photonic Technology and Industrial Policy* (Albany: State University of New York Press, 1992), chap. 4.
50. Lewis M. Branscomb, "Targeting Critical Technologies," in *Empowering Technology: Implementing a U.S. Strategy*, ed. Lewis M. Branscomb, 36–63 (Cambridge, MA: MIT Press, 1993). See also Linda R. Cohen and Roger G. Noll, *The Technology Pork Barrel* (Washington, DC: Brookings Institution, 1991).
51. U.S. National Science Foundation, *Science and Engineering Indicators, 1996*, app. 4, table 4–6.
52. Lewis M. Branscomb and George Parker, "Funding Civilian and Dual-Use Industrial Technology," in *Empowering Technology: Implementing a U.S. Strategy*, ed. Lewis M. Branscomb, 64–102 (Cambridge, MA: MIT Press, 1993).
53. U.S. Council of Economic Advisors, *Economic Report of the President, 1984* (Washington, DC: GPO, 1984), 106.
54. Stockman, *The Triumph of Politics*, 115.
55. George A. Keyworth, "The Role of Science in a New Era of Competition," *Science*, 13 August 1982, 607.
56. Sylvia Ostrey and Richard R. Nelson, *Techno-Nationalism and Techno-Globalism: Conflict and Cooperation* (Washington, DC: Brookings Institution, 1995), especially chap. 2; Dickson, *The New Politics of Science*, especially chap. 1; Branscomb, "Targeting Critical Technologies."
57. Stockman, *The Triumph of Politics*, 114–115. Actually, the demise of Carter's synfuels program was a little more protracted. See Linda R. Cohen and Roger G. Noll, "Synthetic Fuels from Coal," in *The Technology Pork Barrel*, ed. Linda R. Cohen and Roger G. Noll, 259–321 (Washington, DC: Brookings Institution, 1991); and Alfred A. Marcus, *Controversial Issues in Energy Policy* (Newbury Park; Sage, 1992).
58. Smith, *American Science Policy since World War II*, 137–138.
59. U.S. National Science Foundation, *Science and Engineering Indicators, 1996*, app. 4, table 4-6. See also U.S. Office of Science and Technology Policy, *Annual Science and Technology Report*, 108–109; U.S. Congress, Office of

Technology Assessment, *Federally Funded Research: Decision for a Decade* (Washington, DC: OTA, 1991), 73.

60. U.S. National Science Foundation, *Science and Engineering Indicators, 1996*, app. 4, tables 4-5 and 4-6.

61. See U.S. Congress, Office of Technology Assessment, *Starpower: The U.S. and the International Quest for Fusion Energy* (Washington, DC: OTA, 1987), 51.

62. Stockman, for example, was strongly opposed to Clinch River. See *The Triumph of Politics*, 144–146.

63. Linda R. Cohen and Roger G. Noll, "The Clinch River Breeder Reactor," in *The Technology Pork Barrel*, ed. Linda R. Cohen and Roger G. Noll, 217–258 (Washington, DC: Brookings Institution, 1991).

64. Mark Crawford, "Reagan Okays the Supercollider," *Science*, 6 February 1987, 625; and W. D. Kay, "Particle Physics and the Superconducting Super Collider: The Collaboration That Wasn't," in *International Collaboration in Science and Technology among Advanced Industrial Societies*, ed. Vicki L. Golich and W. D. Kay (London: Routledge, forthcoming).

65. See Smith, *American Science Policy since World War II*, 140.

66. Sternberg, *Photonic Technology and Industrial Policy*, 78.

67. Robert B. Reich, "Behold! We Have an Industrial Policy," *New York Times*, 22 May 1988, 29.

68. U.S. Office of Science and Technology Policy, *Annual Science and Technology Report to the Congress, 1982*.

69. Robin Herman, *Fusion: The Search for Endless Energy* (New York: Cambridge University Press, 1990). See also Smith, *American Science Policy since World War II*, 130.

70. "State of the Union Message, January 23, 1983," *Public Papers of the Presidents: Ronald Reagan, 1983*, vol. 1 (Washington, DC: GPO, 1984), 116 (hereafter *Public Papers, 1983*).

71. See Dickson, *The New Politics of Science*, chap. 1.

72. George A. Keyworth, "Federal R&D: Not an Entitlement," *Science*, 18 February 1983, 4586.

73. Reich, "Behold! We Have an Industrial Policy."

74. "Reagan on Aerospace, Aviation," *Aerospace Daily*, 30 September 1980, 167.

75. Peter Duigan and Alvin Rabushka, eds., *The United States in the 1980s* (Reading, MA: Addison-Wesley, 1980). See especially Edward Teller, "Technology: The Imbalance of Power," 281–318.

76. Lou Cannon, "Reagan," in *The Pursuit of the Presidency, 1980*, ed. Richard Harwood, 53–72 (New York: Berkley Books, 1980).

77. Reported in Dickson, *New Politics of Science*, 15.

78. Bruce Murray, *Journey Into Space: The First Thirty Years of Space Exploration* (New York: Norton, 1989), 207fn.

79. Robert Reinhold, "Reagan's Policy on Space Program Still Uncertain," *New York Times*, 9 April 1981, A1, B16.

80. Ibid.

81. See Smith, *American Science Policy since World War II*, 126–127; Jerome B. Wiesner, "The Rise and Fall of the President's Science Advisory Committee." in *Science and Technology Advice to the President, Congress, and the Judiciary*, ed. William T. Golden, 372–384 (New York: Pergamon, 1988).

82. William J. Broad, *Teller's War: The Top-Secret Story behind the Star Wars Deception* (New York: Simon and Schuster, 1992), 104; and Robert Reinhold, "Los Alamos Physicist May Get Post as Science Advisor in the White House," *New York Times*, 8 May 1981, A20.

83. Smith, *American Science Policy since World War II*, 125.

84. Murray, *Journey Into Space*, 210–215. See also U.S. Executive Office of the President, *America's New Beginning: A Program for Economic Recovery* (Washington, DC: Executive Office of the President, 1981), chap. 6, 35–36.

85. W. D. Kay, "Where No Nation Has Gone Before: Domestic Politics and the First International Space Science Mission," *Journal of Policy History* 5.3 (1993): 435–452; and Joan Johnson-Freese, "Canceling the U.S. Solar-Polar Spacecraft," *Space Policy* 3 (1987): 24–37.

86. Smith, *American Science Policy since World War II*, 131

87. Kay, "Where No Nation Has Gone Before," 452fn.

88. In fact, Keyworth himself would later say that his connection to Teller was the primary (if not the only) reason he got the job in the first place). See Broad, *Teller's War*, 104.

89. Smith, *American Science Policy since World War II*, 126–127.

90. Ibid., 126.

91. Howard E. McCurdy, *The Space Station Decision: Incremental Politics and Technological Choice*, Johns Hopkins New Series on NASA History (Baltimore: Johns Hopkins University Press, 1990), 35–36.

92. Interview with James M. Beggs by Howard E. McCurdy, in McCurdy, *The Space Station Decision*, 36.

93. Broad, *Teller's War*, 107.

94. Memo for Martin Anderson, Edwin Harper, Verne Orr, Richard Darman, George A. Keyworth, James Beggs, Hans Mark, and William Schneider from Allan J. Lenz (NSC Staff Director), 17 July 1981, in RRL.

95. "National Space Policy," National Security Decision Directive 42 (4 July 1982). For the complete (unclassified) text, see Christopher Simpson, *National Security Directives of the Reagan and Bush Administrations: The Declassified History of U.S. Political and Military Policy, 1981–1991* (Boulder, CO: Westview, 1995), 136–143 (classified version) and 144–150 (unclassified version). All references are to the latter. Page numbers refer to the Simpson collection.

96. Ibid., 144.

97. U.S. President's Science Advisory Committee, *Introduction to Outer Space: An Explanatory Statement* (Washington, DC: GPO, 1958), 1–2.

98. National Security Council, *Preliminary U.S. Policy on Outer Space*, NSC 5814/1, 18 August 1958, in NASA History Office.

99. "National Space Policy," NSDD 42, 144.

100. Ibid., 144; emphasis added.

101. Ibid., 145.

102. "National Space Strategy," National Security Study Directive 13-82, 15 December 1982, in RRL.

103. The sole exception to this was communication satellites, which had been privately owned and operated since the early 1960s.

104. Quoted in Kathy Sawyer, "Commercialization of Space: Riding on Government Guarantees," *Washington Post*, 1 March 1988, A17.

105. "National Space Policy," National Security Decision Directive 42, 145.

106. Ibid., 147.

107. Ibid., 148.

108. "National Space Strategy," National Security Study Directive 13–82, 15 December 1982, in RRL.

109. "Issue Paper: United States Space Program," White House Office of Policy Development, in RRL.

110. See "President Meets with Space Industry Leaders," *Defense Daily* 129.4 (1983): 1 and Craig Covault, "Reagan Briefed on Space Station," *Aviation Week and Space Technology*, 8 August 1983, 16–18.

111. "Space Commercialization Meeting," memo with agenda, participants, and outline of policy issues, 13 August 1984, in John M. Logsdon, ed., *Exploring the Unknown: Selected Documents in the History of the U.S. Civil Space Program*, vol. 3, *Using Space* (Washington, DC: NASA, 1998), 498–501.

112. Reported in McCurdy, *The Space Station Decision*, 180. See also Covault, "Reagan Briefed on Space Station."

113. "State of the Union Address," *Public Papers of the Presidents: Ronald Reagan, 1984* (Washington, DC: GPO, 1985), 90 (hereafter *Public Papers, 1984*).

114. "National Commercial Space Initiatives (Radio Address to the Nation 21 July 1984)," *Public Papers, 1984*, 1057–1058.

115. "National Space Club Luncheon, 29 March 1985," *Public Papers of the Presidents: Ronald Reagan, 1985*, vol. 1 (Washington, DC: GPO, 1986), 365.

116. "Address to the Nation on Defense and National Security, 23 March 1983," *Public Papers, 1983*, 1:442.

117. Broad, *Teller's War*, chap. 3; and Edward Teller, "Technology: The Imbalance of Power," in *The United States in the 1980s*, Peter Duigan and Alvin Rabushka, (Reading, MA: Addison-Wesley, 1980). For a brief technical discussion of strategic defense technologies, see Alexander Flax, "Ballistic Missile Defense: Concepts and History," in *Weapons in Space*, ed. Franklin A. Long, Donald Hafner, and Jeffrey Boutwell, 33–52 (New York: Norton, 1986).

118. This visit was also significant in that it was when Reagan first got to know Edward Teller, the physicist who would play a major role in advancing SDI. See the oral history interview with Edward Teller, 6 July 1987, in RRL; and also Broad, *Teller's War*, 97–98.

119. Broad, *Teller's War*, 100–101.

120. "1980 Republican Platform Text," *Congressional Quarterly Almanac* (1980): 59-B.

121. Ronald Reagan interview with Lou Cannon, reported in Cannon, *President Reagan*, 320.

122. See, for example, oral history interview, with George A. Keyworth, 28 September 1987, in RRL, 1; and Williams, "The Reagan Administration and Defence Policy," 213.

123. Williams, "The Reagan Administration and Defence Policy," 331–332; and Stockman, *The Triumph of Politics*, 367.

124. On the Joint Chiefs of Staff, for example, see Cannon, *President Reagan*, 330; and Broad, *Teller's War*, 125. On Keyworth, see Broad, *Teller's War*, 131–132. On Secretary of Defense Casper Weinberger, see Cannon, *President Reagan*, 328–329.

125. Keyworth oral history interview, 34–35.

126. See Eric Schmitt, "Now, After $36 Billion Run, Coming Soon: 'Star Wars II,'" *New York Times*, 7 February 1995, A20; and Colin Norman, "SDI Heads for Fiscal Crash," *Science*, 16 March 1990, 1283–1285.

127. See W. Henry Lambright and Dianne Rahm. "Presidential Management of Technology," in *Technology and Politics*, ed. Michael E. Kraft and Norman J. Vig, 81–97 (Durham, NC: Duke University Press, 1988), and "Ronald Reagan and Space Policy," *Policy Studies Journal* 17.3 (1989): 513–527.

128. Paul B. Stares, *The Militarization of Space: U.S. Policy, 1945–1984* (Ithaca, NY: Cornell University Press), 1985, app. 2, table 2.

129. Ibid., chap. 9.

130. "National Space Policy," NSDD 42, 148.

131. Stares, *The Militarization of Space*, 218–219.

132. Colin Norman, "Army, Air Force Eye SDI Spinoffs," *Science*, 16 March 1990, 1284.

133. Stares, *The Militarization of Space*, app. 1, table 2.

134. U.S. National Aeronautics and Space Administration, *Aeronautics and Space Report of the President: Fiscal Year 1992 Activities* (Washington, DC: NASA 1993), app. E-1.

135. "Radio Address to the Nation on Commercial Space Initiatives, July 21, 1984," *Public Papers, 1984*, 1071.

136. White House, "Fact Sheet: National Space Strategy," 15 August 1984. Complete text (and commentary) can be found in Simpson, *National Security Directives of the Reagan and Bush Administrations*.

137. U.S. Congress, House of Representatives, *Space Commercialization: Hearings before the Subcommittee on Space Science and Applications of the Committee on Science and Technology*, 98th Cong., 1st sess., 1983, 60–61.

138. *Congressional Record*, 1984, S-13888. See also Kim G. Yelton, "Evolution, Organization, and Implementation of the Commercial Space Launch Act and Amendments of 1988," *Journal of Law and Technology* 4 (1989): 117–137.

139. Executive Order 12465, "Commercial Expendable Launch Vehicle Activities," *Federal Register* 49.40 (1984): 7211–7212.

140. Public Law 98–575, 1984.

141. Public Law, 100-657, 1988.

142. U.S. Department of Transportation, Office of Commercial Space Transportation, *Fifth Annual Report* (Washington, DC: DOT, 1990).

143. U.S. Congress, Congressional Research Service, *U.S. Commercial Space Activities* (Washington, DC: CRS, 1992).

144. With the exception of those payloads that must be launched from the shuttle.

145. Pamela E. Mack and Ray A. Williamson, "Observing the Earth from Space," in Logsdon, ed., *Exploring the Unknown*, 3:162.

146. For a complete history and introduction to the technology, see Pamela E. Mack, *Viewing the Earth: The Social Construction of the Landsat Satellite System* (Cambridge, MA: MIT Press, 1990).

147. Public Law 98-365. For a complete text see, "Land Remote-Sensing Act of 1984," in Logsdon, ed., *Exploring the Unknown*, 3:329–344.

148. Mack and Williamson, "Observing the Earth from Space."

149. John J. Egan, "Conducting Business and Scientific Experiments in Space," in *Space: National Programs and International Cooperation*, ed. Wayne C. Thompson and Steven W. Guerrier, 135–142, 140 (Boulder, CO: Westview, 1989).

150. Joel S. Greenberg, "The Commercial Launch Industry: Will It Fly on Its Own?" *Aerospace America* (May 1992): 32–35, 41. See also Kathy Sawyer, "Commercialization of Space: Riding on Government Guarantees," *Washington Post*, 1 March 1988, A17.

151. For more on this style of organization, see Hill and Williams, "The Reagan Presidency: Style and Substance," particularly 18; Michael Turner, "The Reagan White House, the Cabinet, and the Bureaucracy," in *Reagan's First Years*, ed. John D. Lees and Michael Turner, 39–67 (Manchester, UK: Manchester University Press, 1988); Shirley Anne Warshaw, "The Reagan Experience with Cabinet Government," in *Ronald Reagan's America*, vol. 2, ed. Eric J. Schmertz, Natalie Datlof, and Alexei Ugrinsky, 717–737 (Westport, CT: Greenwood, 1997); and Deborah Hart Strober and Gerald S. Strober, *Reagan: The Man and His Presidency* (Boston: Houghton Mifflin, 1998), 100–102.

152. National Space Policy," NSDD 42, 149–150. The secretary of transportation was added to this group in 1983, following Reagan's decision to give DOT the responsibility for regulating commercial space launches.

153. Memo to William P. Clark from Craig L. Fuller, 26 June 1982, in RRL.

154. See also W. D. Kay, "Space Policy Redefined: Ronald Reagan and the Commercialization of Space," *Journal of Business and Economic History* 27.1 (1998): 237–247.

155. Cabinet Council Working Group on Space Commercialization, "Potential National Commercial Space Initiatives," June 1984, in RRL.

156. As will be seen in the next chapter, for example, SIG (Space) was not carried over to the Bush administration, and President Clinton chose to relegate

overall responsibility for space policy to the National Science and Technology Council, an organization within OSTP.

157. In fact, it appears that there was serious thought to giving DOC, rather than DOT, responsibility for regulating commercial space launch activity. A November 16, 1983, decision memo to the president from Craig Fuller (RRL) gave Reagan the option of selecting either department for this role. Reportedly, one of the reasons that DOT was chosen was because Secretary Elizabeth Dole promised the president that she could carry out the mission without a budget increase. See "Mrs. Dole Is Given New Space Duties," *New York Times*, 19 November 1983, 13.

158. It should be noted, however, that this particular organization has not always had a comfortable existence with the nation's other space agencies. See, for example, W. D. Kay, *Can Democracies Fly in Space? The Challenge of Revitalizing the U.S. Space Program* (Westport, CT: Praeger, 1995), 81; and Joan Johnson-Freese and Roger Handberg, *Space, the Dormant Frontier: Changing the Paradigm for the 21st Century*, (Westport, CT: Praeger, 1997), 184. As will be seen in the next chapter, it did play a role in the 1988–1989 controversy involving a private space station.

159. Thomas Karas, *The New High Ground: Strategies and Weapons of Space-Age War* (New York: Simon and Schuster, 1983), particularly chap. 1.

160. "National Aeronautics and Space Administration, Remarks at the 25th Anniversary Celebration," *Weekly Compilation of Presidential Documents*, 19 October 1983, 1462–1463.

161. "State of the Union Address," *Public Papers, 1984*, 90. See also "Radio Address to the Nation on the Space Program, 28 January 1984," *Public Papers, 1984*, 108, where Reagan stated that "NASA . . . will be taking a number of initiatives to promote private sector investment" in space activities.

162. "National Space Policy," NSDD 42, 146. At the time, it should be noted, it was believed that virtually all other launch vehicles would be phased out in favor of the shuttle. Thus, this directive effectively gave NASA total responsible for private sector access to space.

163. "Overall Schedule, Cabinet Council Working Group on Space Commercialization," 23 April 1984, in RRL.

164. Cabinet Council Working Group on Space Commercialization, "Potential National Commercial Space Initiatives," in RRL.

165. Public Law 98-361, as amended 1984.

166. National Aeronautics and Space Administration, *NASA Commercial Space Policy*, October 1984 (copy at Space Business Archive).

167. See, for example, Letter from James M. Beggs to Robert C. McFarlane, 17 July 1984, in RRL.

168. "Centers for the Commercial Development of Space Named," *NASA News*, 22 August 1985, 85–120.

169. U.S. National Aeronautics and Space Administration, Office of Commercial Programs, *Centers for the Commercial Development of Space* (Washington, DC: NASA, n.d.).

170. "NASA Agrees to Fly Commercial Spacehab Flights," *Space News Roundup*, 26 August 1988, 4.

171. See "Spacehab Unveils First Module, Predicts Four Flights Annually," *Aviation Week and Space Technology*, 11 May 1992, 59.

172. For a brief history of NASA's early attempts to develop a station, see McCurdy, *The Space Station Decision*, chap. 2.

173. Ibid., 39.

174. See note 111.

175. Letter to the President from Congressmen Jerry Lewis, Manuel Lujan, Bill Lowery, Newt Gingrich, Rod Chandler, and Robert S. Walker, 29 August 1983, in RRL.

176. Cabinet Council Working Group on Space Commercialization, "Potential National Commercial Space Initiatives," in RRL.

177. McCurdy, *The Space Station Decision*, 184.

178. U.S. National Aeronautics and Space Administration, Office of the Space Station, *Space Station Freedom: A Foothold on the Future* (Washington, DC: NASA, 1988).

179. And just as Wernher von Braun and other German scientists and engineers had been willing to work for the Nazis.

180. See Linda T. Krug, *Presidential Perspectives on Space Exploration: Guiding Metaphors from Eisenhower to Bush* (New York: Praeger, 1991), chap. 4.

181. "State of the Union Address," *Public Papers, 1984*, 90.

182. McCurdy, *The Space Station Decision*, 190.

Chapter 8

1. On this last point, see John M. Logsdon, "The Space Shuttle Program: A Policy Failure?" *Science*, 30 May 1986, 1099–1105. It is worth noting in passing that Soviet space proponents had managed to make similar bargains with their political masters; for example, Nikita Khrushchev's support of Sergei Korolev, not because of a desire to explore space but because of the belief that it would give the USSR an advantage in the Cold War. See James Hanford, *Korolev: How One Man Masterminded the Soviet Drive to Beat America to the Moon* (New York: Wiley, 1997); James E. Oberg, *Red Star in Orbit: The Inside Story of Soviet Failures and Triumphs in Space* (New York: Random House, 1981), particularly chaps. 2, 3, 4; and William P. Barry, "How the Space Race Began: The Origins of the Soviet Space Program" (paper presented at the meeting of the Society of the History of the Federal Government, College Park, MD, 3–4 April 1997).

2. For an overview of these and other "anti-NASA" arguments, see W. D. Kay, *Can Democracies Fly in Space? The Challenge of Revitalizing the U.S. Space Program* (Westport, CT: Praeger, 1995), particularly chap. 2.

3. With the exception, of course, of national defense, which (as seen in the preceding chapter) was considered a legitimate function of government, but was also specifically assigned to the Department of Defense by the original National Aeronautics and Space Act that created NASA in the first place.

NOTES TO CHAPTER 8

4. See E. S. Savas, *Privatization: The Key to Better Government* (Chatham, NJ: Chatham House, 1987), 146–147.

5. Milton R. Copulos, "The Perils of a NASA Space Monopoly," *Heritage Foundation Executive Memorandum* 109 (20 June 1984).

6. See note from David Stockman, Director of the Office of Management and Budget to James Beggs, NASA Administrator, 14 June 1982. Copy in Space Business Archive (hereafter SBA).

7. Stuart M. Butler, "An Amtrak for Space? No, Thanks." *Heritage Foundation Executive Memorandum* 19 (11 April 1983).

8. Letter from Representative Ed Zschau to Republican Members of Congress, 14 June 1984, in SBA.

9. Letter from Representative Ed Zschau to President Ronald Reagan, 20 June 1984, in SBA.

10. See Memorandum for Robert McFarlane from Gilbert D. Rye, 19 June 1984. Copy in Ronald Reagan Presidential Library (hereafter RRL). The OMB policy is also mentioned in Letter from Ed Zschau to Republican Members of Congress, 14 June 1984, in RRL.

11. Letter from James M. Beggs to Robert C. McFarlane, 17 July 1984, in RRL.

12. Memorandum for Elizabeth H. Dole from Robert C. McFarlane, 21 June 1984, in RRL.

13. "National Security Decision Directive 144: National Space Strategy (15 August 1984)," in Christopher Simpson, *National Security Directives of the Reagan and Bush Administrations: The Declassified History of U.S. Political and Military Policy, 1981–1991* (Boulder, CO: Westview, 1995), 416–421.

14. "National Security Decision Directive 181: Shuttle Launch Pricing for Foreign and Commercial Customers (30 July 1985)," in Christopher Simpson, *National Security Directives of the Reagan and Bush Administrations: The Declassified History of U.S. Political and Military Policy, 1981–1991* (Boulder, CO: Westview, 1995), 578.

15. Interview with Jenna Dorn (first director of the Office of Commercial Space Transportation), 20 June 1997 (in NASA History Office).

16. For an overview of the ISF story through 1988, see Eliot Marshall, "Space Stations in Lobbyland," Air and Space (December 1988/January 1989): 54–61.

17. Letter to Larry Herbolsheimer, White House, from Maxime A. Faget, 1 March 1984, in RRL.

18. These reports are described in an unsigned memorandum from NASA, Office of the Administrator, to Lillian Trippett, Committee on Science Space and Technology, U.S. House of Representatives, 9 February 1988, in SBA. See also "ISF—A Commercial Space Venture That Raises Administration Dust," *Defense Daily*, 27 January 1988, S1–S4.

19. Lillian Trippett, "White Paper on Industrial Space Facility," Memorandum for Robert Roe, Committee on Science Space and Technology, U.S. House of Representatives, 19 February 1988, 2, in SBA.

20. Ibid.

21. The company also approached the Department of Defense. See "ISF—A Commercial Space Venture," S–2. A 1987 House of Representatives budget authorization report for DOD "encouraged" the department to "examine specific payload requirements that could be met by the ISF" (see Memorandum from Lillian Trippett to Chairman Roe, 4 November 1987, in SBA, 5). There is no indication, however, that DOD ever seriously pursued the matter.

22. Trippett, "White Paper on Industrial Space Facility," 2. See also Kathy Sawyer, "Commercialization of Space: Riding on Government Guarantees," *Washington Post*, 1 March 1988, A17.

23. See Howard E. McCurdy, *The Space Station Decision: Incremental Politics and Technological Choice*, Johns Hopkins New Series on NASA History (Baltimore: Johns Hopkins University Press, 1990), especially chaps. 20 and 23.

24, Quoted in ibid., 213.

25. An excerpt of the continuing resolution can be found as attachment 6 in Trippett, "White Paper on Industrial Space Facility." See also Marshall, "Space Stations in Lobbyland."

26. See, for example, Marshall, "Space Stations in Lobbyland"; "ISF—A Commercial Space Venture That Raises Administration Dust"; and "NASA Use of Industrial Space Platform Ordered by White House Policy Group," *Aviation Week and Space Technology*, 11 January 1988, 4.

27. Letter from James C. Fletcher, NASA Administrator, to Representative Jamie Whitten, Chair of the House Appropriations Committee, 6 January 1988, 1, in SBA.

28. Letter from Senator William Proxmire, Congressman Edward F. Boland, and Congressman Bill Green to James C. Fletcher, 21 January 1988, 1, in SBA.

29. Ibid., 5.

30. Ibid., 1–2.

31. Sawyer, "Commercialization of Space: Riding on Government Guarantees."

32. Marshall, "Space Stations in Lobbyland."

33. Letter from James C. Fletcher to James A. Baker III, Secretary of the Treasury, 31 December 1987, 1, in SBA.

34. See "NASA Use of Industrial Space Platform Ordered by White House Policy Group."

35. White House, Office of the Press Secretary, "The President's Space Policy and Commercial Space Initiative to Begin the Next Century, 11 February 1988," 2, in RRL; and "Fact Sheet: Presidential Directive on National Space Policy" in Christopher Simpson, *National Security Directives of the Reagan and Bush Administrations: The Declassified History of U.S. Political and Military Policy, 1981–1991* (Boulder, CO: Westview, 1995), 856–866. See also William J. Broad, "White House Sets New Space Policy," *New York Times*, 12 February 1988, A36.

36. White House, Office of the Press Secretary, "The President's Space Policy and Commercial Space Initiative," 2.

37. Memorandum from Nancy Risque to Howard Baker, 29 March 1989, in RRL. See also Memorandum for the Economic Policy Council from Eugene J. McAllister, Executive Secretary, 8 March 1988, in RRL.

38. See, for example, a letter from Senators Ernest Hollings and Donald W. Riegle, Jr. to Secretary of Transportation James H. Burnley, 16 March 1988, in RRL.

39. See Andrew Lawler, "Space for Rent?" *Space World* (September 1988): 15–19.

40. William J. Broad, "Senators Opposed on Space Station," *New York Times*, 2 May 1988, B9.

41. National Academy of Sciences, National Research Council, Aeronautics and Space Engineering Board, Commission on Engineering and Technical Systems, *Report of the Committee on a Commercially Developed Space Facility* (Washington, DC: NRC, 1989). See also Kathy Sawyer, "Commercial Space Laboratory Not Needed, Expert Panel Says," *Washington Post*, 12 April 1989, A2; and John Noble Wilford, "U.S. Urged Not to Back Private Space Station," *New York Times*, 12 April 1989, D24.

42. See, for example, Lawler, "Space for Rent?"; "ISF—A Commercial Space Venture That Raises Administration Dust"; and "NASA Use of Industrial Space Platform Ordered by White House Policy Group."

43. Indeed, some of SII's competitors, such as Spacehab (see previous chapter), publicly complained that the administration's policies regarding ISF constituted an unfair advantage. See "Beggs Warns CDSF Threatens Spacehab," *Defense Daily*, 12 April 1988, 245.

44. T. F. Rogers, "Commercial Space Activities: A Disturbing Development" (paper presented at the meeting of the IEEE Aerospace R&D Committee, 23 June 1989) in SBA, particularly 6.

45. Ibid., 1.

46. Quoted in Marshall, "Space Stations in Lobbyland."

47. White House, Office of the Press Secretary, "Fact Sheet: Presidential Directive on National Space Policy," 856.

48. Ibid., 857–858.

49. U.S. President's Commission on the Space Shuttle *Challenger* Accident, *Report* (Washington, DC: GPO, 1986). See also the commentary in Trudy E. Bell and Karl Esch, "The Fatal Flaw in Flight 51-L," *IEEE Spectrum* 24 (1987): 36–51; Frederick F. Lighthall, "Launching the Space Shuttle Challenger: Disciplinary Deficiencies in the Analysis of Engineering Data," *IEEE Transactions on Engineering Management* 38.1 (1991): 63–74; and John L. Hunsucker and Japhet S. Law, "Disaster on Flight 51-L: An IE Perspective on the Challenger Accident," *Industrial Management* 28.5 (1986): 8–13.

50. See, for example, Gregg Easterbrook, "The Spruce Goose of Outer Space," *Washington Monthly* (April 1980): 32–48.

51. Alex Roland, "The Shuttle: Triumph or Turkey?" *Discover* 6 (1985): 29–49.

52. See (among many others) ibid.; Easterbrook, "The Spruce Goose of Outer Space" and "The Case against NASA," *New Republic*, 8 July 1991, 18–24;

Diane Vaughn, *The Challenger Launch Decision* (Chicago: University of Chicago, 1996); Richard Hutton, *The Cosmic Chase* (New York: Mentor, 1981), particularly chap. 7; John M. Logsdon, "The Space Shuttle Program: A Policy Failure?"; and Thomas H. Johnson, "The Natural History of the Space Shuttle," *Technology in Society* 10 (1988): 417–424.

53. "Remarks to Employees of the National Aeronautics and Space Administration on the First Anniversary of the Explosion of the Space Shuttle Challenger," *Public Papers of the President: Ronald Reagan, 1987*, vol. 1 (Washington, DC: GPO, 1988), 80.

54. "Statement on the Building of a Fourth Shuttle Orbiter and the Future of the U.S. Space Program," *Public Papers of the President: Ronald Reagan, 1986*, vol. 2 (Washington, DC: GPO, 1987), 1111.

55. Of course, as has already been seen, it was also supporting a commercial competitor (from NASA's point of view) at the same time.

56. Memorandum from James A. Baker III to the Economic Policy Council, 6 October 1986, in RRL.

57. See U.S. Department of Transportation, Office of Commercial Space Transportation, *Fifth Annual Report* (Washington, DC: DOT, 1990); and "FAA-Licensed Launches Set Record in 1998," *AST News Bulletin* 3.4 (1999): 1.

58. "U.S. Leads World in Commercial Market," *AST News Bulletin* 3.4 (1999): 1.

59. White House, Office of the Press Secretary, "The President's Space Policy and Commercial Space Initiative," 3.

60. The only real exceptions are those satellites and probes that require use of the shuttle. By the end of the 1990s, the selection of commercial launch vehicles for its missions had become one of NASA's most important procurement items. See Warren Ferster, "NASA Payload Sizes Give Delta Launcher and Edge," *Space News*, 22 March 1999, 1, 19.

61. See Joel S. Greenberg, "The Commercial Launch Industry: Will It Fly on Its Own?" *Aerospace America* (May 1992): 32–35, 41.

62. See interview with Jenna Dorn, 20 June 1997.

63. White Houose, Office of the Press Secretary, "The President's Space Policy and Commercial Space Initiative," 5.

64. Although, as previously noted, this was eventually shifted to private launch providers.

65. See Joseph C. Anselmo, "Shuttle Fails to Replace ELVs, but Pentagon Eyes It Again," *Aviation Week and Space Technology*, 8 April 1996, 52–53.

66. The European Space Agency and Japan had insisted on a prohibition of military activity aboard the station as a condition of joining the project. See McCurdy, *The Space Station Decision*, 104–105.

67. See Roger D. Launius, *NASA: A History of the U.S. Civil Space Program* (Malabar, FL: Krieger, 1994), 114.

68. Roger Pielke, "Space Shuttle Value Open to Interpretation," *Aviation Week and Space Technology*, 26 July 1993, 57–58.

69. McCurdy, *The Space Station Decision*, 230–235.

70. See U.S. Congress, House of Representatives, *Proposed Space Station Freedom Program Revisions: Hearings before Committee on Science, Space, and Technology*, 101st Cong., 1st sess., 31 October 1989; and *Station Design Flaws: Hearings before the Committee on Space, Science, and Technology*, 101st Cong., 2d sess., 29 March 1990. See also Christopher Anderson, "U.S. Space Station Needs Another Redesign," *Nature*, 29 March 1990, 367; and U.S. Congress, General Accounting Office, *Space Station: NASA's Search for Design, Cost, and Schedule Stability Continues* (Washington, DC: GAO, 1991).

71. William J. Broad, "How the $8 Billion Space Station Became a $120 Billion Showpiece," *New York Times*, 10 June 1990, A1, A30, and "Higher Estimate for Maintaining Station in Space," *New York Times*, 11 July 1990, A1, B6.

72. Eliot Marshall, "Space Station Science: Up in the Air," *Science*, 1 December 1989, 110–112; Christopher Joyce, "NASA's Incredible Shrinking Space Station," *New Scientist*, 17 November 1990, 14; Eric J. Lerner, "Space Station Changes Its Course," *Aerospace America* (January 1991): 12–15; and U.S. Congress, General Accounting Office, *Questions Remain on the Costs, Uses, and Risks of the Redesigned Space Station* (Washington, DC: GAO, 1991. See also W. D. Kay, "Democracy and Super Technologies: The Politics of the Space Shuttle and Space Station *Freedom*," *Science, Technology, and Human Values* 19.2 (1994): 131–151.

73. See, for example, Gregg Easterbrook, "NASA's Space Station Zero," *Newsweek*, 11 April 1994, 30–33.

74. See, for example, Hans Mark, *The Space Station: A Personal Journey* (Durham, NC: Duke University Press, 1987).

75. It is worth noting, however, that as its capabilities diminished, it was becoming less able even to support this goal. NASA's original plans to develop new technologies for "closed-loop" life-support systems and more sophisticated on-board computers (essential for extended missions outside near-earth orbit) had to be shelved due to their high cost. See McCurdy, *The Space Station Decision*, 114–117.

76. Craig Covault, "Hubble Team Makes Emergency Plans as Space Telescope Fails Focus Testing," *Aviation Week and Space Technology*, 2 July 1990, 23–24, and "NASA Fights Image, Technical Problems on Hubble, Shuttle," *Aviation Week and Space Technology*, 9 July 1990, 16–18.

77. Colin Macilwain, "Did Last-Minute Change Cause Loss of $1-Billion Mars Probe?" *Nature*, 2 September 1993, 3. See also John Travis, "Mars Observer's Costly Solitude," *Science*, 3 September 1993, 1264–1267; and John Travis and John Cohen, "The Sounds of Silence: Chronology of Despair," *Science*, 3 September 1993, 1265.

78. Frederic Smoller, "The View from the Future," *American Heritage* (November 1992): 70–72.

79. U.S. Congress, Congressional Budget Office, *Preliminary Analysis of NASA Commercialization Initiatives: Staff Memorandum* (Washington, DC: CBO, 1989), in SBA, 10.

80. U.S. National Aeronautics and Space Administration, *Aeronautics and Space Report of the President, Fiscal Year 1995 Activities* (Washington, DC: NASA, 1996), app. A-1.

81. Ibid., app. E-1B.

82. Tom Wolfe, *The Right Stuff* (New York: Bantam Books, 1979).

83. This is, of course, a relative statement. As seen in previous chapters, *Vanguard* and other such misadventures did generate a great deal of criticism.

84. And, it should be noted, the "service" was one originally seen as directly related to national security.

85. "National Security Directive 30: National Space Policy (2 November 1989)," in Christopher Simpson, *National Security Directives of the Reagan and Bush Administrations: The Declassified History of U.S. Political and Military Policy, 1981–1991* (Boulder, CO: Westview, 1995); 914–928. The Space Council was established by Executive Order 12675.

86. "Space Council Bypasses NASA to Get Industry Moon–Mars Ideas," *Space News*, 11 December 1989, 4.

87. See Kathy Sawyer, "Quayle, Citing Kennedy Legacy, Vows to Fight to Save Space Station Program," *Washington Post*, 17 May 1991, A7; Helen Gavaghan, "Cut-Price Space Station Runs Out of Steam," *New Scientist*, 23 March 1991, 13; Patricia A. Gilmartin, "Bush Administration Rallies Support for Space Station as Crucial Vote Nears," *Aviation Week and Space Technology*, 27 May 1991, 25–26; and Peter Aldhous, "Space Station Back on Track," *Nature*, 13 June 1991, 507.

88. Kathy Sawyer, "Bush Urges Commitment to New Space Exploration," *Washington Post*, 21 July 1989, 4. See also "Space Exploration Initiative," NSPD-6, 13 March 1992.

89. See Launius, *NASA*, 124.

90. "National Security Directive 30," 920.

91. See "Fact Sheet: Presidential Directive on National Space Policy," 860.

92. Lyn Ragsdale, "Politics, Not Science: The U.S. Space Program in the Reagan and Bush Years," in *Spaceflight and the Myth of Presidential Leadership*, ed. Roger D. Launius and Howard E. McCurdy, 133–171, 161 (Urbana: University of Illinois, 1997).

93. Quoted in Roger D. Launius and Howard E. McCurdy, introduction to *Spaceflight and the Myth of Presidential Leadership*, ed. Roger D. Launius and Howard E. McCurdy, 1–14, 8 (Urbana: University of Illinois, 1997).

94. See ibid., 1–14. For a specific example, see Bruce Murray, "Civilian Space: In Search of Presidential Goals," *Issues in Science and Technology* 2.3 (1986): 25–37.

95. Indeed, one of the similarities with Kennedy borders on the eerie. Both elaborated on their initial announcements in speeches at Texas universities: Kennedy at Rice in 1962, Bush at Texas A&M in 1990. See Roger D. Launius and Howard E. McCurdy, epilogue to *Spaceflight and the Myth of Presidential Leadership*, ed. Roger D. Launius and Howard E. McCurdy, 221–250, 229 (Urbana: University of Illinois, 1997).

96. Launius, *NASA*, 124.

97. U.S. Congress, House of Representatives, *President's Request for Mars and Lunar Missions: Hearings before the Subcommittee on Space Science and Applications of the Committee on Science, Space, and Technology*, 101st Cong., 1st sess., 26 September 1990.

98. See Clifford Kraus, "House Halts Work on Supercollider," *New York Times*, 25 June 1993, A18; Jeffrey Mervis and Karen Fox, "A Senate Victory Would Turn the Tide after House Defeat," *Science*, 16 July 1993, 288–289; and David Ritson, "Demise of the Texas Supercollider," *Nature*, 16 December 1993, 607–610

99. Patricia A. Gilmartin, "House Kills Funding for Mars/Moon Effort," *Aviation Week and Space Technology*, 2 July 1990, 28.

100. D. Allan Bromley, "Science, Scientists, and the Science Budget," *Issues in Science and Technology* 9.1 (1992): 44.

101. This may, of course, be more of a comment on the Bush administration itself. During both the 1988 and 1992 presidential campaigns, Bush faced criticism that he lacked both "vision" (particularly in 1988) and a coherent domestic policy agenda (in 1992).

102. The latter was still under development when Goldin was appointed, but was too far along to cancel.

103. U. S. National Aeronautics and Space Administration, *NASA Pocket Statistics, 1996* (Washington, DC: NASA, 1996), app. C, tables C-28 and C-29.

104. Seth Borenstein, "NASA Probe to Mars Reported Lost," *Boston Globe*, 24 September 1999, A3.

105. Deborah Zabarenko, "NASA Faulted in Mars Failure," *Boston Globe*, 29 March 2000, A3.

106. U.S. Office of Science and Technology Policy, National Science and Technology Council, "Fact Sheet: National Space Policy," 19 September 1996, 5–6.

107. U.S. Office of Science and Technology Policy, National Science and Technology Council, "National Space Transportation Policy," NSTC-4, 5 August 1995, sec. III (1) (a) and sec. IV (1) (c).

108. Ibid., sec. III (2) (d).

109. Ibid., sec. IV (2) (a).

110. Peter B. deSelding, "Government, Private Spending Nearly Equal," *Space News*, 19 April, 1999, 3, 20. For fiscal year 2001, however, NASA was slated to receive a modest increase, its first in eight years.

111. And even that received nothing like the personal presidential attention given to the space station or SEI.

112. U.S. National Aeronautics and Space Administration, *Aeronautics and Space Report of the President*. To be fair, it should be noted that at least part of the reduction in NASA's budget was due to Goldin's downsizing policies.

113. The most well-known of these during the Reagan–Bush years was SEMATECH, a government–industry corporation set up in 1987 to conduct research

on advanced semiconductor technology. See David P. Angel, *Restructuring for Innovation: The Remaking of the U.S. Semiconductor Industry* (New York: Guilford, 1994); and Peter Grindley, David C. Mowery, and Brian Silverman, "SEMATECH and Collaborative Research: Lessons in the Design of High-Technology Consortia," *Journal of Policy Analysis and Management* 13 (1994): 723–758.

114. This was, for example, clearly the case with SEMATECH. See Richard D. Bingham, *Industrial Policy American Style: From Hamilton to HDTV* (Armonk, NY: Sharpe, 1998), chap. 6; and Lewis M. Branscomb, "The National Technology Policy Debate," in *Empowering Technology: Implementing a U.S. Strategy*, ed. Lewis M. Branscomb, 1–35 (Cambridge, MA: MIT Press, 1993).

115. U.S. Executive Office of the President, *Technology for America's Economic Growth: A New Direction to Build Economic Strength* (Washington, DC: White House, 1993). See also Clinton–Gore Campaign, *Technology: The Engine for Economic Growth: A National Technology Policy for America* (Little Rock, AR: Clinton–Gore Campaign, 1992).

116. Charles V. Shank, "Recharging the Energy Laboratories," *Physics Today* (July 1994): 46.

117. Denis F. Cioffi, "Clinton Philosophy Transforms NIST into 'Partner for Industry,'" *Physics Today* (November 1994): 75–77.

118. The Republicans also took control of the Senate for the first time in ten years.

119. Sensenbrenner succeeded Walker as chair of the full Science Committee following the latter's retirement in 1996. His successor on the Space and Aeronautics Subcommittee, Dana Rohrbacher, was an official in the Reagan administration, and has been highly critical of what he regards as the Clinton administration's lack of interest in the space program (see, for example, his remarks in "Space Shots," *Space News*, 8 March 1999, 14).

120. Peter Bond, "Can Russian Skills Revive the Space Station?" *New Scientist*, 30 October 1993, 34–36; Warren E. Leary, "With Russian Aid, Better Space Lab," *New York Times*, 5 November 1993, A20; and Jeffrey M. Lenorovitz, "Russia May Hold the Space Station Key," *Aviation Week and Space Technology*, 23 August 1993, 22–24.

121. Jeffrey M. Lenorovitz and Boris Rybak, "Feeble Russian Economy Hinders Space Efforts," *Aviation Week and Space Technology*, 15 March 1993, 91–92; "Russian Space Program Seeks Needed Funding," *Aviation Week and Space Technology*, 23 August 1993, 24; Craig Covault, "95,000 Russian Layoffs, Launch Breakdown Feared," *Aviation Week and Space Technology*, 15 November 1993, 27; and Yuri Kolesnikov, "Russian Cosmonautics in Difficult Times," *Spaceflight* 35 (1993): 366–367. Not only did the Russian program have less money than before, but it had additional expenses as well. The Soviet Union had built its launch complex in Kazakhstan, which, following the USSR's collapse, had become an independent country. As a result, the Russian government has been forced to pay a fee to use the facility. See Fred Hiatt, "Russia Leases Back Cosmodrome," *Washington Post*, 29 March 1994, A14; and Daniel Sneider, "Kazakhstan, Russia Reach Agreement over Baikonur," *Space News*, 4–10 April 1984, 12.

122. See, for example, Simon Saradzhyan, "Soyuz Maker Demands Space Agency Pay Its Bills" *Space News*, 13 September 1999, 8.

123. Simon Saradzhyan and Brian Berger, "NASA Predicts Longer Service Module Delay," *Space News*, 22 February 1999, 4, 28.

124. James F. Sensenbrenner, "Be Realistic about Russia's Role in the Space Station," *Space News*, 10 May 1999, 24; and Paula Shaki, "U.S. Bill Could Halt Payments to Russia," *Space News*, 14 June 1999, 6,; and "Lawmakers Grill Goldin on Russian ISS Work Delays," *Space News*, 28 February 2000, 38.

Chapter 9

1. See (among many others) Ann Markusen and Joel Yudken, *Dismantling the Cold War Economy* (New York: Basic Books, 1992); Center for Strategic and International Studies, Steering Committee on Security and Technology, *Integrating Commercial and Military Technologies for National Strength: An Agenda for Change* (Washington, DC: CSIS, 1991); Carnegie Commission on Science, Technology, and Government, *Technology and Economic Performance: Organizing the Executive Branch for a Stronger National Technology Base* (New York: Carnegie Commission, 1991); Lewis M. Branscomb, ed., *Empowering Technology: Implementing a U.S. Strategy* (Cambridge, MA: MIT Press, 1993); Jacques Gansler, *Defense Conversion: Transformation of the Arsenal of Democracy* (Cambridge, MA: MIT Press, 1995); and U.S. Congress, General Accounting Office, *The Department of Energy: National Laboratories Need Clearer Missions and Better Management* (Washington, DC: GAO, 1995).

2. Admittedly, the present problem is much larger in scale and scope. The $36 billion decline in spending for defense procurement between 1985 and 1993 represents a decrease, in real terms, of more than 50 percent. Such reductions could have had much more substantial economic consequences than the NASA cutbacks 25 years earlier.

3. DOE inherited the laboratories from the Atomic Energy Commission in 1977.

4. See Department of Energy, *Department of Energy, 1977–1994* (Washington, DC: DOE, 1994); Bernard Cole, "DOE Labs: Models for Tech Transfer," *IEEE Spectrum* (December 1992): 53–57; and Charles V. Shank, "Recharging the Energy Laboratories," *Physics Today* (July 1994): 42–48. It should be noted that the $6 billion does not include the DOD laboratories, defense-related university research, and the large number of private firms existing solely on defense contracts.

5. DOE, *Department of Energy, 1977–1994*, 100.

6. Quoted in Lewis M. Branscomb, "National Laboratories: The Search for New Missions and New Structures," *Empowering Technology: Implementing a U.S. Strategy*, ed. Lewis M. Branscomb, 103–134, 110 (Cambridge, MA: MIT Press (1993).

7. John Burgess, "Bombs into Bulldozers," *Washington Post*, 23 August 1992, 1.

8. Gansler, *Defense Conversion*; Markusen and Yudken, *Dismantling the Cold War Economy*.

9. Jay Stowsky and Burgess Laird, "Conversion to Competitiveness: Making the Most of the National Labs," *American Prospect* (Fall 1992): 91–98; and Branscomb, "National Laboratories."

10. Shank, "Recharging the Energy Laboratories," 45.

11. See Scott McCartney, "With Cold War Over, Los Alamos Seeks New Way of Doing Business," *Wall Street Journal*, 15 July 1993, 1.

12. Ray Kidder, "Conversion of U.S. Nuclear Weapons Laboratories," in *Conversion of Military R&D*, ed. Judith Reppy, 201–214 (New York: St. Martin's, 1998).

13. Shank, "Recharging the Energy Laboratories."

14. U.S. Congress, House of Representatives, *Electric Vehicles and Advanced Battery R&D: Hearings before the Subcommittee on Energy, Committee on Science, Space, and Technology*, 103rd Cong., 2nd sess., 30 June 1994.

15. Bob Davis, "White House and Auto Makers Prepare Joint Effort to Triple Fuel Efficiency," *Wall Street Journal*, 29 September 1993, 1. See also U.S. Department of Commerce, Partnership for a New Generation of Vehicles, *Program Plan* (Washington, DC: DOC, 1994), and *Inventions Needed for PNGV* (Washington, DC: DOC, 1995); and Daniel Sperling, "Updating Automotive Research," *Issues in Science and Technology* (Spring 2002): 85–89.

16. U.S. Department of Energy, Office of Utility Technologies, *Superconductivity Program for Electric Power Systems: FY 1994–1998 Multi-Year Plan* (Washington, DC: DOE, 1994).

17. A more detailed discussion—with an assessment—of these and other projects can be found in Richard D. Bingham, with Maria Papadakis, *Industrial Policy American Style: From Hamilton to HDTV* (Armonk, NY: Sharpe, 1998), chap. 7.

18. Jeff Bingaman and Bobby R. Inman, "Broadening Horizons for Defense R&D," *Issues in Science and Technology* 9.1 (1992): 80–85.

19. Bruce Berkowitz, "Can Defense Research Revive U.S. Industry?" *Issues in Science and Technology* 9.2 (1992/1993): 74. See also Lewis M. Branscomb and George Parker, "Funding Civilian and Dual-Use Industrial Technology," in *Empowering Technology: Implementing a U.S. Strategy*, ed. Lewis M. Branscomb, 64–102, 65–67 (Cambridge, MA: MIT Press, 1993).

20. It is not yet clear how this perspective has been changed by the war on terror and increased concerns over homeland security.

21. Cole, "DOE Labs."

22. Kenneth Flamm, *Creating the Computer: Government, Industry, and High Technology* (Washington, DC: Brookings Institution, 1988).

23. Lewis M. Branscomb, "The National Technology Policy Debate," in *Empowering Technology: Implementing a U.S. Strategy*, ed. Lewis M. Branscomb, 103–134, 110, 34 (Cambridge, MA: MIT Press, 1993). It should be noted that this is quite different from the more familiar notion of a "spin-off." In the past, any civilian application that was "spun off" of some military technology was generally regarded by the defense establishment as little more than an after-the-fact "bonus" (see Berkowitz, "Can Defense Research Revive U.S.

Industry?"). The dual-use doctrine, however, treats commercial applications as a primary objective.

24. Public Law 99-502.
25. Quoted in Branscomb, "National Laboratories," 103.
26. U.S. Executive Office of the President, Council on Competitiveness, *Industry as a Customer of the Federal Laboratories* (Washington, DC: Council on Competitiveness, 1992). It should be noted, however, that in endorsing this policy, the council was engaging in a bit of redefinition of its own. The basic thrust of the report is that industrial modernization and the further development of the high-tech private sector are both an important element in military readiness. In other words, the involvement of government laboratories in the civilian economy—which under most circumstances economic conservatives would wish to avoid (see chapter 7)—is regarded as acceptable here due to national security considerations.
27. John M. Logsdon, "The Evolution of U.S. Space Policy and Plans," in *Exploring the Unknown: Selected Documents in the History of the U.S. Civil Space Program*, vol. 1, *Organizing for Exploration*, ed. John M. Logsdon, 377–393, 386 (Washington, DC: NASA, 1995).
28. U.S. Congress, House of Representatives, *Toward the Endless Frontier: History of the Committee on Science and Technology, 1959–1979* (Washington, DC: GPO, 1990), 312
29. Quoted in ibid., 309.
30. See John M. Logsdon, "The Space Shuttle Program: A Policy Failure?" *Science*, 30 May 1986, 1099–1105.
31. It could also be argued that, in its earliest discussions of a space station program, NASA considered this same "multiple-use" approach, promising that the system could serve a variety of users. Its use for defense purposes, however, was soon dropped. See Howard E. McCurdy, *The Space Station Decision: Incremental Politics and Technological Choice*, Johns Hopkins New Series on NASA History (Baltimore: Johns Hopkins University Press, 1990).
32. W. Henry Lambright, *Powering Apollo: James E. Webb of NASA*, Johns Hopkins New Series in NASA History (Baltimore: Johns Hopkins University Press, 1990), 139.
33. This is described in McCurdy, *The Space Station Decision*.
34. This was largely due to the delays in developing the shuttle, as well as NASA's failure to account fully for the effects of atmospheric "drag."
35. Lewis M. Branscomb, "Research and Innovation Policy: A Framework for Research-Based Industrial Policy" (paper presented at the Conference on U.S.–E.U. Technology Policy, Atlanta, GA, 9–10 April 1999), 2. See also Ivan Eland, "Can the Pentagon Be Run Like a Business?" *Issues in Science and Technology* (Spring 2002): 78–84.
36. Such as the National Ignition Facility at Lawrence Livermore, which was highly controversial even before running up large cost overruns. See "National Ignition Facility: DOE set to Double the Price of the Troubled Laser Project," *Science*, 5 May 2000, 782; "U.S. National Ignition Facility Hits Crisis,"

Nature, 9 September 1999, 101; and Tom Zamora Collina, "The National Ignition Facility: Buyer Beware," *Technology Review*, Feb.–March 1997, 37–40.

37. To take only one example (among many), the Sandia National Laboratory, once heavily involved in the development and testing of nuclear weapons, has more recently been working on building a safer handgun. See "Sandia Researchers Shoot for 'Smart Gun.'" *DOE This Month* 18.3 (1995): 12.

38. Patricia A. Gilmartin, "House Kills Funding for Mars/Moon Effort." *Aviation Week and Space Technology*, 2 July 1990, 28.

39. A call that has been intensified by the possible discovery of water below the Martian surface. See Sharon Begley and Erika Check, "NASA: Mars Is All Wet," *Newsweek*, 3 July 2000, 48–50.

40. Holman Jenkins Jr., "Big Question on Mars: What Are We Waiting For?" *Wall Street Journal*, 28 June 2000, A23.

41. Ibid.; Jeffrey Kluger, "Will We Live on Mars?" *Time*, 10 April 2000, 60–63; and Robert M. Zubrin, with Richard Wagner, *The Case for Mars: The Plan to Settle the Red Planet and Why We Must* (New York: Free Press, 1996).

42. Jenkins, "Big Question on Mars."

43. See Logsdon, "The Evolution of U.S. Space Policy and Plans," 382; and Lambright, *Powering Apollo*. See also Logsdon's "James E. Webb: A Dominant Force in 20th Century Public Administration," *Public Administration Review* 53.2 (1993): 95–99.

44. Such as W. D. Kay, "Democracy and Super Technologies: The Politics of the Space Shuttle and Space Station *Freedom*," *Science, Technology, and Human Values* 19.2 (1994): 131–151, and *Can Democracies Fly in Space? The Challenge of Revitalizing the U.S. Space Program* (Westport, CT: Praeger, 1995).

45. Whether this would have been a good or a bad thing is another matter.

46. There would, of course, have to be some understanding that these assignments would be revisited (and perhaps revised) as the economic and political conditions surrounding spaceflight changed.

47. Launius, *NASA*, 124; and Ragsdale, "Politics, Not Science," 161.

48. Launius, *NASA*.

49. The cartoon is reprinted in Walter A. McDougall, . . . *the Heavens and the Earth: A Political History of the Space Age* (New York: Basic Books, 1985), 167. For a more detailed discussion of the interservice rivalries during this period, see the classic John M. Logsdon, *The Decision to Go to the Moon: Project Apollo and the National Interest* (Chicago: University of Chicago Press, 1970).

50. For more on this phenomenon, see Christopher J. Bosso and W. D. Kay, "Advocacy Coalitions and National Space Policy," in *Space Politics and Policy: An Evolutionary Perspective*, ed. Eligar Sadeh, 43–57 (Amsterdam: Kluwer Academic, 2002).

51. What makes this situation especially odd is that a very large aerospace company could carry out both tasks simultaneously. See ibid.

52. Actually, one of the other implications of this discussion is how much longer it will be accurate even to use the term *community*.

53. Of course, as chapter 8 demonstrates, such questions were finally raised, but not until after a surprisingly long time (particularly given the Reagan administration's views on the topic).

54. For a discussion of this issue from the standpoint of another field, see Andrea L. Bonnicksen, *Crafting a Cloning Policy: From Dolly to Stem Cells* (Washington, DC: Georgetown University Press, 2002), chap. 1.

55. See Gregg Easterbrook, "Long Shot," *Atlantic Monthly* (May 2003): 65–75; "Rocket Propulsion Companies Explore New Strategies to Cope with Weak Market," *Space News*, 15 July 2002, 17; and Stan Crock, "Call it the Wrong Stuff," *Business Week*, 14 October 2002, 148–150.

56. "The Pendulum Swings," *Space News*, 24 February 2003, 22. See also Peter B. deSelding, "Satellite Manufacturers Look Hopefully toward 2003," *Space News*, 13 January 2003, 8; and Brian Berger, "Group Forecasts Diminished Commercial Satellite Sales," *Space News*, 24 February 2003, 25.

57. Andy Pasztor and Anne Marie Squeto, "Lockheed's Atlas Rocket Faces Slack Demand in Sector Slump," *Wall Street Journal*, 23 August 2002, A2; Warren Ferster, "Delta 4 Gives Boeing Fresh Start: Maiden Flight Lifts Spirits in a Difficult Launch Market," *Space News*, 25 November 2002, 1, 4; and Jason Bates, "Lack of Satellite Orders Bodes Badly for Launch Providers," *Space News*, 13 January 2003, 8.

58. Quoted in Brad Stone, "Bezos in Space," *Newsweek*, 5 May 2003, 52.

59. For a general discussion, see Kim S. Cameron, Robert I. Sutton, and David A. Whetten, eds., *Readings in Organizational Decline* (Cambridge, MA: Ballinger, 1988); and R. A. D'Aveni, "The Aftermath of Organizational Decline: A Longitudinal Study of the Strategic and Managerial Characteristics of Declining Firms," *Academy of Management Journal* 32 (1989): 577–605.

60. Crock, "Call it the Wrong Stuff."

61. Ibid.

62. Easterbrook, "Long Shot"; Peter B. deSelding, "Satellite Operators Set Sights on Military Market," *Space News*, 2 December 2002, 19, "Satellite Operators Turning to Military Market for Growth," *Space News*, 3 February 2003, 20, and "Satellite Operators Expect Increased Demand from U.S," *Space News*, 10 March 2003, 7; and Jason Bates, "U.S. Commits to Purchase Commercial Satellite Imagery," *Space News*, 20 January 2003, 1, 3.

63. "September 11," *Space News*, 9 September 2002, 14. See also the cautionary passages in Stone, "Bezos in Space."

64. Incidentally, it was quite common, in the days after September 11, to hear and read of the attacks being compared to the Japanese bombing of Pearl Harbor in 1941. Similarly, there were officials in 1957 worrying that *Sputnik* was "another Pearl Harbor."

65. See, for example, Jim Dawson, "National Labs Focus on Tools against Terrorism in Wake of Airliner and Anthrax Attacks," *Physics Today* (January 2002): 19–22.

66. See Gareth Cook, "A Window on Osama's World," *Boston Globe*, 25 September 2001, C1, C3; and "Brave New World of Computer Imagery," *Space News*, 13 January 2003, 22.

67. Jeremy Singer, "U.S. War on Terror Fuels Work on New Military Space Concepts," *Space News*, 16 September 2002, 17, 18, and "U.S. Army Policy Spells Out Critical Space Capabilities," *Space News*, 9 June 2003, 21.

68. U.S. Congress, General Accounting Office, *Military Space Operations: Common Problems and Their Effects on Satellite and Related Acquisitions* (Washington, DC: GAO, 2003). See also "Big Increase Projected for U.S. Military Space Spending," *Space News*, 13 January 2003, 4; and Jeremy Singer, "Pentagon Seeks $600 Million Space Budget Boost," *Space News*, 3 February 2003, 6.

69. A fact acknowledged by even some of the agency's supporters. See, for example, "September 11."

70. Holman W. Jenkins, Jr., "What the Shuttle Commission Won't Say," *Wall Street Journal*, 23 July 2003, A15.

71. Alex Beam, "Columbia: A 'Normal Accident'?" *Boston Globe*, 4 February 2003, E1; Marica Dunn, "Discovery of Molten Aluminum Boosts Shuttle Breakup Theory," *Boston Globe*, 5 March 2003, A2; Brian Berger, "Working Scenario Pegs Wing Breach in Columbia Disaster," *Space News*, 28 April 2003, 8, and "Columbia Hypothesis Released; NASA Names Shuttle Manager," *Space News*, 12 May 2003, 15; and R. Jeffrey Smith, "Columbia Board Reports NASA Faults," *Boston Globe*, 13 July 2003, A5.

72. Berger, "Columbia Hypothesis Released; NASA Names Shuttle Manager"; and "NASA Shakes Up Shuttle Program," *Boston Globe*, 3 July 2003, A4.

73. xfind citesx.

74. Kembra J. Dunham and Kris Maher, "NASA Faces Shortfall in Funding, Skilled Applicants," *Wall Street Journal*, 4 February 2003, B1, B4.

75. Even if there was the political will (not to mention the financial resources) to do so, the general feeling of the space community is that the present shuttle technology is seriously out of date.

76. See, for example, U.S. Congress, General Accounting Office, *Space Transportation: Challenges Facing NASA's Space Launch Initiative* (Washington, DC: GAO, 2002).

77. Andrew C. Revkin, "Now the Space Station: Grieving, Imperiled," *New York Times*, 4 February 2003, D1, D4; and Brian Berger, "NASA Has Few Options for ISS during Shuttle Grounding," *Space News*, 10 February 2003, 7.

78. Hiawatha Bray, "Should We Be Up There at All?" *Boston Globe*, 4 February 2003, C1, C4; Warren Ferster, "Like Challenger, Columbia Prompts Soul Searching," *Space News*, 10 February 2003, 1, 3; Eric Pianin and Kathy Sawyer, "Shuttle Future Is Under Debate," *Boston Globe*, 12 May 2003, A4; and Sharon Begley, Anne Marie Squeo, and Guy Chazan, "The Hard Question for NASA: Can the Space Station Survive?" *Wall Street Journal*, 5 February 2003, A1, A4.

Index

Adams, Sherman, 43, 44
Advanced Research Projects Agency, (ARPA) 54,55
American Association for the Advancement of Science, 121
American Rocket Society, 28, 33
Amtrak, 119
anti-satellite weapons, 131–2
Apollo Applications Project. *See Skylab*
Apollo fire, 7, 99, 104,
Apollo program, 4, 6, 10, 19, 64, 76, 106, 139, 153, 155, 157, 168, 169, 172, 176
   criticisms of 77, 79, 81–82, 84, 96–97
Apollo XI, 85, 91–92, 130, 156, 157
Atomic Energy Commission (AEC), 13, 14, 17,55, 57

ballistic missile defense. *See* Strategic Defense Initiative
Baker, James, 140, 149
Bay of Pigs, 72, 81
Beggs, James M., 126, 139, 140, 146
biotechnology, 14–15
Branscomb, Lewis, 169
British Interplanetary Society, 28
Brooks, Overton, 6, 69
Brown, George, 167, 168
Bureau of Alcohol, Tobacco, and Firearms (ATF), 133
Bureau of the Budget, 81–82, 92, 94, 104, 107

Bush, George H. W., 4, 67, 150, 156, 157, 160, 161, 163, 168, 172, 173
Bush, George W., 22, 167, 176, 177, 178
Bush, Vannevar, 32, 33, 77

Cabinet Council on Commerce and Trade, 136–38, 140
Carey, William, 121
Carson, Rachel, 97
Carter, Jimmy, 13, 15, 17, 23, 108, 109, 110, 118, 122, 123 , 124
Cassini probe, 159
Central Intelligence Agency, 38
Challenger (space shuttle), 5, 7, 99, 135, 145, 146, 147, 151–53, 154, 160, 177, 178
China, People's Republic of, 9, 33
Clarke, Arthur C., 33
Clinch River Breeder Reactor, 123
Clinton, William J., 22, 122, 123, 160–62, 163, 167
Cold War, 5, 18, 21, 44, 46–48, 49,56, 58, 59, 62, 75, 76, 83, 107, 110, 114, 127, 131, 141, 144, 155, 156, 165, 166, 168, 169, 170
   as justification for more public R&D spending, 49–50, 100, 104
Collier's magazine, 34
Columbia (space shuttle), 5, 7, 125, 126, 151, 177, 178
communication satellites, 115–16

Commercially Developed Space Facility.
  See Industrial Space Facility
Conestoga I, 133
Congress, U. S., 5, 9, 14, 21–22, 50–51,
  55–58, 63, 134, 136, 144, 145, 146,
  152, 156, 159, 163, 164, 167, 168,
  171
Cuba, 72, 79

Dallek, Robert, 94
Defense Advanced Research Projects
  Agency (DARPA), 54, 165, 166, 167,
  169, 170
definition. *See* problem definition
Democratic Party, 5, 43, 50, 61, 65, 160,
  161, 162, 176
Department of Commerce, 137, 149, 150,
  156, 161
Department of Defense (DOD), 6, 42–43,
  48–49, 54, 97, 104–5, 113, 120,
  132, 136, 137, 139, 153, 166, 168,
  178
Department of Energy (DOE), 16, 18–9,
  20, 23, 119, 166
Department of Housing and Urban
  Development (HUD), 7
Department of Transportation (DOT),
  119, 133, 134, 137, 146, 152, 156
Department of Veteran's Affairs 7
Dery, David, 11, 13
Disney, Walt, 34
Dornberger, Walter, 29, 3
Dryden, Hugh, 71
dual use technology, 165, 168
DuBridge, Lee A., 107
Dulles, Allen, 38–39

Echo I, 61
Eisenhower, Dwight D., 33, 34, 35, 36,
  37, 39–40, 41, 54, 55,57, 60, 64, 72,
  73, 119, 127, 155
  criticisms of, 43–44, 48–49, 61, 71, 100
  response to *Sputnik,* 42–44, 48, 50, 176
energy policy, 15, 122
European Space Agency, 162
Explorer I, 42, 49, 61

Federal Aviation Administration, 133
Federal Communications Commission, 133

First Men in the Moon (Wells), 27, 28
Fletcher, James, 149
Fulbright, J. William, 79, 173
Fuller, Craig, 136, 140
fusion energy program, 16, 20, 21, 49–50

Gagarin, Yuri, 68, 69, 71, 72, 73, 86
Galileo probe, 125, 126, 159
Gemini Project, 86, 100, 102, 111
German Society for Spaceship Travel.
  See Verein fur Raumschiffhrt
Gilruth, Robert, 85
Glennan, T. Keith, 57, 59, 62, 64
Goddard, Robert, 27, 28–29
Goldin, Daniel, 158–60
Gorbachev, Mikhail, 120
Greenberg, Daniel S., 50

Hale, Edward Everett, 27
Herblock, 43, 173
high-energy physics, 50
Hitler, Adolf, 30, 31, 120, 143
Hornig, Donald F., 96
House of Representatives
  Appropriations Committee, 49
  Joint Committee on Atomic Energy,
    54
  Legislative Reorganization Act, 54
  Science and Astronautics Committee,
    44, 49, 51,71, 104, 163, 168
Hubble space telescope, 8

Industrial policy, 119
Industrial Space Facility (ISF), 146–50,
  154
intercontinental ballistic missile (ICBM),
  32, 37, 38, 45
International Geophysical Year (IGY),
  35–37
International Space Station (ISS), 4, 19,
  162, 178
  Reagan's approval of, 10, 139–41
issue definition. *See* problem definition
issue entrepreneur, 29

Japan, 32, 162, 167
Jet Propulsion Laboratory, 37
Johnson, Lyndon B. 23, 34, 51–53, 55,57,
  63, 65, 67, 105, 129, 157, 171

and post-Apollo space program, 92–97, 110
as chair of National Aeronautics and Space Council, 83–84
as president, 86, 94, 96, 98, 100
as vice president, 72–73, 77
Johnson Space Center, 85, 93

Kennedy, John F., 9, 10, 6, 63, 71, 76, 77, 100–01, 129, 161
moon landing goal, 71–72, 73–75, 106, 141, 157, 158, 172
proposals for joint spaceflights with USSR, 10, 61–62, 78–81, 84–85
Keyworth, George, 122, 124, 126, 127, 140
Khrushchev, Nikita, 47–48, 68, 80, 85
Killian, James, 33, 34, 39, 45, 60
Kisaowsky, George, 60

Laird, Melvin, 107
LANDSAT, 119, 134
Lawrence Livermore National Laboratory, 166, 169
Lenin, Vladimir, 68
Ley, Willy, 34
Life magazine, 43
Lodge, Henry Cabot, 44
Los Alamos National Laboratory, 126, 166, 167
Lovell, Bernard, 80
Luna I, 42
Luna II, 42
Luna III, 42

Manhattan Project, 32, 34
Manned Orbiting Laboratory, 139
Manned Space Center. *See* Johnson Space Center
Mark, Hans, 139
Mars, 4, 5, 34, 92, 102, 106, 108, 140, 154, 156, 170, 171, 172, 177
Mars Observer probe, 154, 159
Marshall Space Flight Center, 31
Mayo, Robert P., 107
McCormack, John W., 55, 56
McCurdy, Howard, 35, 140
McNamara, Robert S., 73, 74
Meese, Edwin, 25

Mercury program, 4, 59–60, 61, 67–68, 70, 111, 147, 155
Metcalf, Lee, 56
Mondale, Walter, 113
Murray, Bruce, 125

Nader, Ralph, 97
National Academy of Sciences, 28–29, 150
National Advisory Committee on Aeronautics, 29, 55, 122
National Aeronautics and Space Act, 6, 44, 46, 53, 55–56, 57, 62, 72, 76, 136, 137, 172
National Aeronautics and Space Administration (NASA), 114, 115, 155, 160, 164, 165, 168, 170, 172, 174, 175, 176, 178
administrator, 57, 117, 159
and access to space, 135
budget, 86, 87, 92, 95, 108, 111, 113, 126, 148, 161
Centers for the Commercial Development of Space, 138–39, 147
commercial programs, 135, 143–44, 150, 154
criticisms of, 4–5, 144, 145–46, 150, 152, 154
external environment, 7, 44
founding of, 44, 53–59, 55–57, 65
mission statement, 7
science programs, 82, 112, 126, 154
Space Task Group (STG), 107, 108, 112
National Aeronautics and Space Council, 57, 63, 65, 71, 73, 156, 161
National Aerospace Plane (NASP), 4
National Defense Education Act, 54
National Institutes of Health (NIH), 119
National Science Foundation (NSF), 33, 36, 49, 54, 119, 124, 125
National Security Council, 36, 38, 43, 48–49, 57, 60, 107
National Space Club, 130
Naval Research Laboratory, 37–38
New York Times, 29, 98
Newsweek, 110, 117
Nixon, Richard M., 23, 39, 43–44, 107, 108, 109, 110, 112, 143

INDEX

Oak Ridge National Laboratory, 106
Oberth, Herman, 27, 29, 30, 34
Office of Defense Mobilization, 33
Office of Management and Budget (OMB), 81, 107, 111, 122, 146, 152, 156
Office of Science and Technology Policy, 125, 161
Open Skies, 33, 36
Orbiter project, 37, 38

Paine, Thomas O., 107
Peenemünde, 30, 31, 37
Pravda, 68, 86
President's Science Advisory Committee (PSAC), 54, 55
Press, Frank, 108, 109
Privatization, 118, 120, 134
problem definition, 11–9, 24
   and human spaceflight, 67–68, 141
   and information 13–4, 38–39
   and NASA 19, 63, 84–85, 87
   and participation 14–5, 37–38
   and rocketry 29–36, 39
   and satellites 33–34
problem ownership, 12–3
Proxmire, William, 94, 148
Public Health Service, 12

Quarles, Donald, 37, 38, 48–49
Quayle, Dan, 156, 158, 168

Redefinition, 16–7, 33–34, 127
RAND Corporation, 34
Reedy, George, 51, 52
Reagan, Ronald, 9, 10, 15, 116, 117, 124, 125 , 128, 132, 133, 134, 136, 137, 144, 155, 157, 160, 161, 162, 163, 168, 173
   and privatization, 119
   and shuttle pricing, 145
   and space station, 139–41, 144, 147, 152
   energy policy, 122–23
   R&D policies, 121–23
   tax policy, 118
   views on space and government, 126, 129, 152, 158
reconnaissance satellites, 34, 48–49, 176–77

Redstone, 31, 37, 49
Remote sensing, 109
Reorganization, 22–24, 65, 136
Republican Party, 5, 50, 52, 116, 162, 163, 176
research and developmemt policy, 10, 20, 21, 99
Rusk, Dean, 73
Russia, 162, 163

Sandia National Laboratory, 166, 169
Schultze, Charles L., 96–97, 103
Seamans, Robert, 71
Senate
   Aeronautical and Space Sciences Committee, 51, 73
   Appropriations Committee, 103
   Commerce, Science, and Transportation Committee, 51, 150
   Subcommittee on Preparedness, 51
Senior Interagency Group for Space 136, 138, 140, 156, 161
Shepard, Alan B., 70, 74
Skylab, 93, 101, 108, 168
Smithsonian Institution 28
Soviet Union. *See* Union of Soviet Socialist Republics
Soyuz, 101
Space Exploration Initiative (SEI), 4, 156–58, 163, 170, 172
Space Industries Incorporated, 147
space race, 9, 37, 38, 39, 42, 47–48, 73
space shuttle program (see also specific craft), 19, 112–13, 115, 126, 137–38, 143, 158, 160, 168, 169, 178
   pricing, 144–46, 150
space station. *See* International Space Station
Space Transportation System (STS). *See* space shuttle program
Sputnik I, 10, 24, 37, 40, 41, 44, 52,55, 63, 65, 68, 70–71, 72, 77, 104, 109, 136, 173, 176
   and Soviet propaganda, 38, 45–47, 86–87
   psychological impact of, 45, 47–48, 54
Sputnik II, 41, 45

Sputnik III, 45
Stevenson, Adlai., 53
Stockman, David, 122, 140
Strategic Defense Initiative (SDI), 125, 131
Superconducting Super Collider (SSC), 16, 123, 157, 158
Supersonic Transport, 112

Teague, Olin, 84, 93–94, 98
technical lock-in, 20–21, 78, 110–12
Teller, Edward, 124, 126
terrorism, 15, 176–77
Thomas, Albert, 84
Three Mile Island, 16
Truman, Harry S., 38, 52, 53
Tsiolkovsky, Konstantin, 27, 28

U-2 spy plane, 33–34, 36, 39
Union of Soviet Socialist Republics (USSR), 9, 10, 21, 33, 35, 36, 37, 38–39, 41, 42, 45, 47–48, 49–50, 57, 58, 59,61, 73–74, 85, 101, 102, 104, 111, 120, 126, 128, 143, 155, 165, 168
United Nations, 44, 80

V-2 rocket, 30–31, 32
Van Allen radiation belts, 61
Vanguard, 37, 49, 50, 55, 64
Verein fur Raumschiffhrt (VfR), 28, 30

Verity, William, 128, 149
Verne, Jules, 27, 28
Vietnam War, 14, 96, 97, 98, 100
Viking rocket, 37
von Braun, Wernher, 27, 30–32, 34–35, 37–38, 42, 49, 74, 104, 143
Vostok I, 68,70, 72, 82 , 104
Voskhod 1, 86–87
Voskhod 2, 87, 100, 101
Voyage dans la lune, Le (film), 28

Waterman, Alan, 49
Watkins, James, 168
weather satellites, 3, 61, 76, 109
Webb, James, 63, 71, 72, 77, 80, 82, 92–93, 94, 95, 98, 103–04, 110, 160, 169, 171
Weinberg, Alvin, 106
Weinberger, Caspar, 111
Welsh, Edward, 71, 101, 104
Wells, H. G., 27, 28
White, Ed, ix, 87, 99, 102
White Sands, 31
Wiesner, Jerome, 62, 67
Wilson, Charles, 43
Wolfe, Tom, 72, 155
World War II, 13, 29–31
WS-117L, 34, 36

Zubrin, Robert, 170–71